Miniaturized Testing of
ENGINEERING MATERIALS

Advanced Materials Science and Technology

Series Editor
Baldev Raj

Miniaturized Testing of Engineering Materials
V. Karthik, K.V. Kasiviswanathan, and Baldev Raj

Miniaturized Testing of
ENGINEERING
MATERIALS

V. Karthik
K.V. Kasiviswanathan
Baldev Raj

CRC Press
Taylor & Francis Group
Boca Raton London New York

CRC Press is an imprint of the
Taylor & Francis Group, an **informa** business

CRC Press
Taylor & Francis Group
6000 Broken Sound Parkway NW, Suite 300
Boca Raton, FL 33487-2742

First issued in paperback 2020

ISBN 13: 978-0-367-57455-0 (pbk)
ISBN 13: 978-1-4822-6391-6 (hbk)

Library of Congress Cataloging-in-Publication Data

Names: Karthik, V., author. | Kasiviswanathan, K. V. | Raj, Baldev, 1947-
Title: Miniaturized testing of engineering materials / author: V. Karthik,
K.V. Kasiviswanathan, and Baldev Raj.
Description: Boca Raton : Taylor & Francis, CRC Press, 2017. | Series:
Advanced materials science and technology
Identifiers: LCCN 2016011477 | ISBN 9781482263916 (hard cover)
Subjects: LCSH: Materials--Microscopy. | Materials--Testing. | Microchemistry.
Classification: LCC TA417.23 .K37 2017 | DDC 620.1/127--dc23
LC record available at https://lccn.loc.gov/2016011477

Visit the Taylor & Francis Web site at
http://www.taylorandfrancis.com

and the CRC Press Web site at
http://www.crcpress.com

Dedicated to Sri K. V. Kasiviswanathan, coauthor,

who passed away on August 26, 2014.

Contents

Preface

The conventional mechanical tests for evaluating tensile, fatigue, fracture toughness, and creep properties of engineering materials are well established and codified into standards. The standard specimens are typically a few centimeters long and a few millimeters in other dimensions so as to satisfy polycrystalline behavior and to represent bulk material response. Meeting the specimen size requirements for conventional mechanical testing may not always be possible and there is a need to reduce the specimen dimensions in practical situations, such as in characterization of weld joints, coatings, new and exotic materials, life assessment of operating components, radioactive materials, failure analysis, and testing miniaturized devices like microelectromechanical systems (MEMS). Mechanical testing using specimen sizes much smaller than the conventional specimen sizes is broadly referred to as "miniature specimen testing." Historically, the development of this technology originated in the nuclear industry in the early 1980s to cater to the requirements of material development for fission and fusion communities. Driven by scientific interest as well as industrial applications of miniature specimen testing, many advances have taken place across the globe in the last three decades across the length scale from millimeter down to nanometer. In this book, we have focused on the small-scale specimens in the size range of 0.3–3.0 mm and attempted to provide a comprehensive coverage of various test methods of characterizing the mechanical properties of engineering materials. These include scaled-down versions of tensile, impact, and fracture toughness tests as well as the punch and indentation-based nonconventional techniques.

Numerous journal papers on this topic have been published by researchers from various countries and this subject has also been exclusively and extensively dealt with in ASTM series conferences since 1986, as well as in the European groups. Round-robin testing exercises in specific techniques within countries and across various countries have been in vogue to evolve a common code of practice (CoP) so as to make small specimen tests acceptable to designers, plant operators, and regulatory authorities. With numerous methodologies of small specimen testing emerging, maturing, and finding spin-off applications in materials science and technology, there was a compelling need for a comprehensive book on this subject matter that could serve as unified resource material for the benefit of both materials researchers and industry professionals. Toward this objective, this book is a compilation of various small specimen test techniques, giving the reader a flavor of the evolution, innovations, and novelty in experimental and numerical studies, correlation methodologies, and size effects of each of the small specimen test techniques.

The first chapter introduces the various conventional mechanical test methods and the specimen sizes they employ to the reader and emphasizes the need for employing miniaturized specimens in applications where conventional mechanical tests are, practically, not possible. The second chapter deals with various miniaturized test methods for the determination of tensile properties and associated flow curve of material. The third chapter is devoted to miniaturized specimen test methods that have evolved for the fatigue and fracture toughness determination. In the fourth chapter, specimen size effects and their influence on the measured mechanical properties are presented. Various factors, such as number of grains, constraints, and deformation/failure mechanisms that give rise to potential size effects, are analyzed and presented for each of the test techniques. The fifth and final chapter deals with applications of miniature specimen testing technology through a few case studies of relevance to both materials scientists and industry professionals. Typical examples will span the broader areas of applications in both nuclear and other fields of materials science and engineering and are thematically covered as (1) nuclear (irradiated components); (2) power plant structures; (3) narrow zones such as weld joints, coatings, etc.; and (4) diverse applications such as for characterizing nanomaterials, biomaterials, etc.

The findings and results of pioneering research work of numerous investigators, together with research activity in our organization, have been packaged with appropriate citations. We sincerely acknowledge these contributions and also record our appreciation of Dr. Gagandeep Singh and his team for their dedicated efforts to realize our dream of writing this work.

We dedicate this book to Sri K. V. Kasiviswanathan, one of our coauthors who is no longer with us.

We invite constructive criticism from the readers, which we are sure will help us to revise this book for future editions.

V. Karthik
Baldev Raj

Authors

Dr. V. Karthik is currently a senior scientist at Indira Gandhi Center for Atomic Research (IGCAR), Kalpakkam, India. He holds a postgraduate degree in physics from the University of Hyderabad and an engineering degree in metallurgy from the Indian Institute of Metals. Working for the past 21 years in the metallurgy and materials group of IGCAR, his research interests include characterization of mechanical behavior of materials using conventional and small sized specimens and characterization of irradiated fuel and structural materials of fast reactors and analysis of their in-reactor performance. Dr. Karthik has steered the research program on the indigenous development of novel characterization methods that employ small specimens for mechanical property evaluation, which culminated in a doctoral thesis on this topic. He was a recipient of the young engineer award (2006) from the Department of Atomic Energy (DAE) and has over 40 publications in peer-reviewed journals, as well as international and national conference proceedings.

K. V. Kasiviswanathan served the Department of Atomic Energy, India, for more than 40 years and held various R&D, project management, and implementation positions in IGCAR, Kalpakkam. His areas of specialization and interest included postirradiation examination of fast reactor fuel and structural materials, development of irradiation testing techniques and equipment, development of miniature specimen testing techniques, development of various in-service inspection (ISI) systems, development of robotics and remote handling equipment, and failure analysis of engineering components. He was responsible for setting up the unique hot cell facility for post irradiation examination (PIE) of plutonium-rich fuel and contributed toward understanding the irradiation performance of indigenously developed unique mixed carbide fuels of a fast breeder test reactor at various stages of its burnup. He was awarded the Indian Nuclear Society (INS) outstanding service award in 2009 for outstanding contributions in the area of nuclear fuel cycle technologies. He had over 150 publications in journals, international and national conference proceedings, and books.

Dr. Baldev Raj, BE, PhD, served the Department of Atomic Energy, India, over a 42-year period until 2011. As distinguished scientist and director, Indira Gandhi Center of Atomic Research, Kalpakkam, he has advanced several challenging technologies, especially those related to the fast breeder test reactor and the prototype fast breeder reactor. Dr. Raj pioneered the application of non destructive evaluation (NDE) for basic research using acoustic and electromagnetic techniques in a variety of materials and components.

He was pivotal for initiating ab initio research and developmental activities on miniature specimen testing at IGCAR. He is also responsible for realizing societal applications of NDE in areas related to cultural heritage and medical diagnosis. He is the author of more than 970 refereed publications, 70 books and special journal volumes, and more than 20 contributions to encyclopedias and handbooks, as well as the owner of 29 patents. He is immediate past president of the International Institute of Welding, and president of the Indian National Academy of Engineering. He assumed responsibilities as the director of the National Institute of Advanced Studies, Bangalore, in September 2014. He is a fellow of all science and engineering academies in India, member of the German Academy of Sciences, honorary member of the International Medical Sciences Academy, member of the International Nuclear Energy Academy, vice president of nondestructive testing (NDT), Academia International, and president-elect of the International Council of Academies of Engineering and Technological Sciences.

1

Introduction

1.1 Materials and Properties

From the Stone Age to the present silicon-driven Information Age, materials have occupied the center stage in the development and advancement of civilization. This is evident from the fact that various ages have been named after materials—for example, the Stone Age, the Bronze Age, and the Iron Age. The domain of materials science and engineering is a very exciting area directed toward understanding why metals and materials behave the way they do, how materials are made, and how new materials with unique properties can be created.

The characteristics of materials are often described in terms of physical, chemical, electrical, and mechanical properties. Physical properties refer to the color, density, melting point specific heat, and heat conductivity, while flammability, corrosion/oxidation resistance, and toxicity are some examples of chemical properties. Electrical properties comprise electrical conductivity, resistance, and magnetic qualities of the material. The mechanical properties relate to elasticity, load-carrying ability, hardness, and wear resistance. The measurement of physical, electrical, and mechanical properties forms an important area in the study of metals and materials.

Mechanical properties reflect the elastic and plastic behavior of a material in response to applied loads, time, temperature, and other conditions. Important mechanical properties are strength, hardness, ductility, and toughness—to name a few. These properties are useful for two primary purposes: (1) for design of engineering structures based on strength or deflections, and (2) to meet material specifications during material processing and fabrication. In the former case, with the knowledge of the mechanical properties, the structural engineer ensures that the load endured by the structure or resulting deformation does not become excessive and lead to fracture. In the latter case, the knowledge of mechanical properties is of interest to materials and metallurgical engineers who are involved in material processing and fabrication. This necessarily involves an understanding of the relationships between the microstructure of materials and their mechanical properties.

1.2 Mechanical Properties and Microstructure

The strength of a material is its ability to withstand an applied stress without failure. The field of strength of materials deals with loads acting on a material and the resulting elastic and plastic deformations. Plastic deformation is permanent, and strength and hardness are measures of a material's resistance to this deformation. A material's strength is dependent on its microstructure. Metallic materials are, in general, polycrystalline containing an aggregate of very small crystals, which are called grains. On a microscopic level, plastic deformation of metals corresponds to the motion of dislocations in response to externally applied shear stress through a process termed slip. Dislocations are defects of the crystal lattice that may be introduced during solidification and or during plastic deformation.

The ability of a metal to plastically deform depends on the ability of dislocations to move by slip processes. However, slip occurs on specific crystallographic planes and in certain directions that depend on the crystal structure of the material. Dislocation motion occurs along the slip system that has the most favorable orientation (i.e., the highest shear stress). Due to random crystallographic orientations of the numerous grains, the direction of slip varies from one grain to another. The atomic disorder within a grain boundary region results in a discontinuity of slip planes from one grain into the other. Grain boundaries thus serve as barriers to dislocation motion, and dislocation accumulation at the grain boundaries contributes to the macroscopic deformation hardening of the material (Figure 1.1). Similarly, lattice strain interactions between impurity atoms (that go into either substitutional or interstitial solid solution) and dislocations result in solid solution strengthening. Strain hardening, which is the enhancement of strength with increased plastic deformation, is caused by an increase in dislocation density and by repulsive strain field interactions of dislocations. The main factors

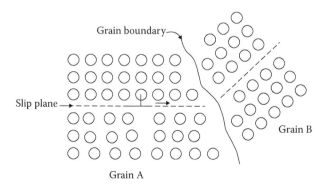

FIGURE 1.1
Schematic showing motion of dislocation and change of slip plane direction across grain boundary.

that determine the strain hardening or work hardening of an alloy are the elementary processes of dislocation motion such as glide and interactions between dislocations or between dislocation and obstacles in the interior of the grains. The underlying mechanisms in deformation behavior of metals and their alloys can thus be understood with the knowledge of the dislocations and the role they play in plastic deformation processes.

1.3 Techniques for Mechanical Property Characterization

Mechanical testing is a general term referring to the broad range of activities involved with the determination of mechanical properties and behavior of materials and structures. The mechanical properties of materials are determined by performing carefully designed laboratory experiments on material component or model specimens that replicate as nearly as possible the service conditions. The laboratory experiments involve applying a stimulus-like force, pressure, displacement, heat, and/or their combinations and measuring the response of the system. The mechanical tests are classified depending upon the nature of the applied load and its duration, as well as the environmental conditions. The load could be tensile, compressive, or shear, and its magnitude may be constant with time, or it may fluctuate continuously. The application of the load may be only a fraction of a second, or it may extend over a period of many years. A brief outline of the fundamental types of mechanical tests that have evolved over the years is presented in the following section.

Tensile and compressive testing are the most fundamental tests to determine strength and ductility properties of common engineering materials. In a tensile test, a machined specimen of typical dimensions (shown in Figure 1.2) is deformed to fracture, with a gradually increasing load that is applied along the specimen axis. The load is applied using a universal testing machine (UTM) on which the specimen is mounted at its ends into the holding grips of the machine (Figure 1.3); the specimen is continuously elongated at a constant rate until failure occurs. The stress–strain plot (Figure 1.4) consists of three distinctly different regions: (1) the elastic region where stress

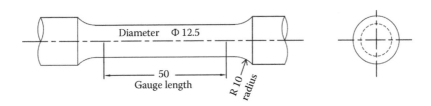

FIGURE 1.2
Sketch of a typical tensile test specimen (all dimensions in millimeters).

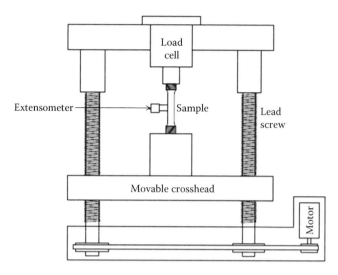

FIGURE 1.3
Schematic of a universal tensile testing machine.

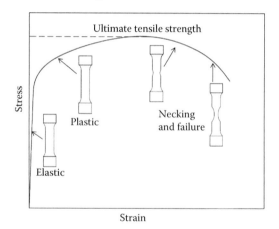

FIGURE 1.4
Stress–strain curve of a polycrystalline material showing the various stages of deformation.

and strain increase linearly up to the proportional limit and removing the load allows the strain to return to zero; (2) the plastic regime beyond the elastic region, where the specimen is said to have yielded and undergoes an increase in stress due to increases in strain (also referred to as strain hardening); and (3) a peak in the load-extension curve signifying onset of plastic instability and decrease in load leading to final failure. The load-elongation data are analyzed to compute the properties such as strengths (yield, ultimate, rupture, etc.), strains (elongation), elastic modulus, Poisson's ratio, and uniaxial stress–strain relationships. A compression test is conducted in a manner similar to the tensile test, except that the force is compressive and the specimen contracts along the direction of the stress. Compressive tests are used when the material is brittle in tension (such as concrete) or when a material's behavior under large and permanent strains is desired, as in metal-forming applications.

Just like the tensile and compression tests, which measure the resistance of a material to plastic flow, the hardness test is an age-old mechanical test for a quick and easy assessment of the deformation characteristics and can be related to the tensile or compressive strength. Though hardness is not a fundamental property of the metal, it is widely employed for determining the suitability of a material for a given application and quick inspection after a particular treatment to which the material has been subjected. Simply stated, hardness is the resistance of a material to permanent deformation. Hardness tests can be static indentation type, rebound type, or scratch type tests. Static hardness tests involve the use of a specifically shaped indenter made of diamond, carbide, or hardened steel pressed into the material with a known force. The hardness values are determined by measuring either the depth of indenter penetration (like in Rockwell hardness) or the size of the resultant imprint such as in Brinell, Knoop, and Vickers hardness tests.

Purely static loading is seldom observed in modern engineering components or structures. By far, the majority of structures involve parts subjected to fluctuating or cyclic loads. Fatigue is a condition whereby a material cracks and fails as a result of cyclic stresses that are typically much lower than that required to cause failure in static loading. Fatigue testing is performed on materials by subjecting a specimen to cyclic stresses that may be axial (tension–compression combinations), flexural (bending), or torsional (twisting). Typical fatigue test machines are shown in Figure 1.5. Most fatigue tests

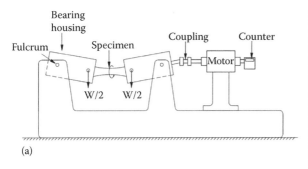

(a)

(b)

FIGURE 1.5
Universal testing machines employed for fatigue testing. (a) Rotating beam fatigue machine and (b) modern fatigue machine.

are conducted at what is referred to as "constant amplitude," which refers to the fact that the maximum and minimum stresses are constant for each cycle of a test (Figure 1.6). In these tests, also called high-cycle fatigue, the stress levels are elastic and below the yield strength of the material. Under such conditions, the material will endure millions of cycles before failure. A plot of constant amplitude stress level (S) versus number of cycles to failure (N_f) is obtained in such tests (Figure 1.7), from which the fatigue limit or endurance limit of the material can be computed. Low-cycle fatigue testing involves straining the specimen to beyond the yield stress, thereby accumulating damage in the specimen and leading to failure in much fewer numbers of cycles.

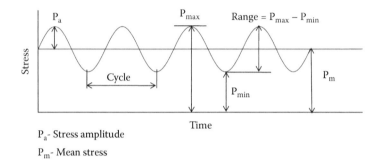

P_a- Stress amplitude

P_m- Mean stress

FIGURE 1.6
Load cycling in a typical high-cycle fatigue test.

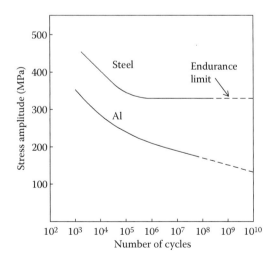

FIGURE 1.7
Typical S–N plot obtained in a high-cycle fatigue test.

The ability of metals to plastically deform is commonly referred to as "ductility." However, failure of metal parts often occurs in a catastrophic manner without prior plastic deformation and is referred to as a brittle fracture. A brittle failure generally occurs at unpredictable levels of stress, by rapid crack propagation. Using the concepts of fracture mechanics, it is possible to determine whether a crack of given length in a material is dangerous at a given stress level. Fracture toughness is the material property that indicates the amount of stress required to propagate a crack from a preexisting flaw. Fracture toughness data not only provide guidelines for selection of materials and design against fracture failures, but also aid in establishing component life spans, along with inspection and maintenance criteria.

One of the approaches to evaluate the fracture toughness is based on the concepts of linear elastic fracture mechanics (LEFM), which relates stress intensity factor or toughness to flaw size (a), component geometry, and applied stress (σ) for an elastic loading condition except for a very small region near the crack tip (Figure 1.8). The property K_{IC}, also called plain strain fracture toughness, characterizes the resistance of a material to fracture in the presence of a sharp crack under tensile loading, where the state of stress near the crack front is triaxial plane strain. The K_{IC} test involves the loading to failure of precracked notched specimens in tension or three-point bending (Figure 1.9) under quasi-static conditions. Fracture toughness is computed using expressions involving load corresponding to a defined increment of crack length, specimen thickness and width, and geometric function relating the compliance of the specimen to the ratio of crack length and width. The test must be validated for meaningful results by ensuring certain size

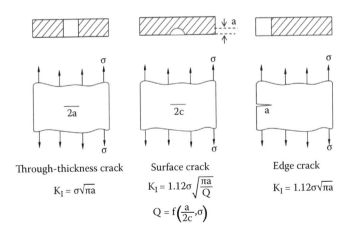

Through-thickness crack

$$K_I = \sigma\sqrt{\pi a}$$

Surface crack

$$K_I = 1.12\sigma\sqrt{\frac{\pi a}{Q}}$$

$$Q = f\left(\frac{a}{2c}, \sigma\right)$$

Edge crack

$$K_I = 1.12\sigma\sqrt{\pi a}$$

FIGURE 1.8
Various crack configurations and stress intensity factors.

FIGURE 1.9
Compact tension (CT) specimen and three-point bend specimen loading for K_{IC} determination.

requirements to ensure plane strain (triaxial constraint) conditions at the crack tip.

For materials that are highly ductile with significant crack tip plasticity, specimen requirements to achieve plain strain conditions are very large and sometimes practically impossible to perform a valid K_{IC} test. For such materials, it is necessary to use nonlinear fracture parameters such as the J-integral and crack tip opening displacement (CTOD) to describe the fracture criterion at the crack tip. Toughness, in general, depends on strength, ductility, temperature, loading rate, and microstructure. An unnotched bar of ductile metal will not fracture under impact loading but will undergo plastic deformation. However, in the presence of a notch, a metal can undergo brittle fracture under impact loading. This phenomenon is more common in body-centered cubic (BCC) and in some hexagonal close-packed metals, which undergo a sharp change from ductile to brittle behavior and loss of toughness across a narrow temperature range called the transition temperature. Notch-impact tests are most widely used to provide information on the resistance of such materials to sudden fracture in presence of a flaw. In these tests, the energy required to produce rupture is measured to determine the relative tendency of brittleness as a function of temperature. The most common tests of this type are the "Charpy V-notch" and "Izod" tests performed using a pendulum type of machine. Specimen sizes and loading concepts are illustrated in Figure 1.10. The notched-bar Charpy impact tests are conducted over a range of temperatures to generate a transition–temperature curve (Figure 1.11) so that the temperature at which the ductile-to-brittle transition takes place in a BCC alloy can be determined. Additional information is obtained if the impact test is instrumented to provide a load-line history of the specimen during the test, from which the energy required for initialing fracture and the energy required for propagating fracture can also be determined. Apart from the Charpy impact tests, another test method that has gained acceptance for determination of nil ductility temperature is the Naval Research Laboratory drop weight test.

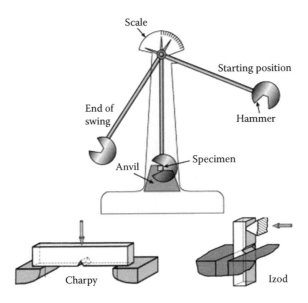

FIGURE 1.10
Specimen configurations and loading systems employed for impact testing.

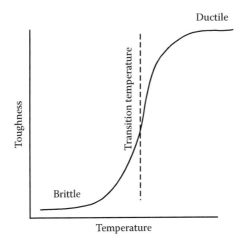

FIGURE 1.11
Typical S-shaped curve obtained in Charpy impact test at various temperatures.

In the transition range, the value of fracture toughness is not only temperature dependent but also scattered even at a fixed temperature. In recent years, research has been focused toward accounting for the scatter, size effects, and temperature dependence of fracture toughness and improving the confidence levels of fracture toughness in the transition regime. This has resulted in the master curve approach, which is a statistical, theoretical, micromechanism-based analysis method for fracture toughness in the ductile-to-brittle transition region. This approach makes use of probability distribution functions to describe the fracture toughness data, which vary from full cleavage at lower temperatures to fully ductile at higher temperatures within the transition zone.

The mechanical strength of metals decreases with increasing temperature and the properties become much more time dependent. Metals subjected to a constant load at temperatures above half the melting temperatures (T_M) undergo "creep," a time-dependent change in dimensions. A creep test involves measuring the dimensional changes accurately at constant high temperature and constant load or stress. A specimen very similar to a tensile specimen is mounted inside a furnace and fitted with a sensitive strain-measuring device for long durations ranging from 2000 to 20,000 h. From the plot of creep strain versus time (Figure 1.12), the minimum secondary creep rate is computed, which is of most interest for high-temperature design and performance. There exist various models of creep laws available for extrapolating creep behavior of materials beyond the test times and predicting the behavior at longer durations.

There are several other methods of testing not described here, such as the torsion test, bend test and those pertaining to corrosion testing, weld joint performance, and some tailored specifically to a product (e.g., fasteners, washers). Each of the mechanical test techniques described in the preceding paragraphs has several forms with variations in the specimen type, rate of loading, stress states, and procedures for data analysis. Readers may refer to Dieter (1990), Callister and Rethwisch (2011), and Kuhn and Medlin (2008) for more on mechanical testing.

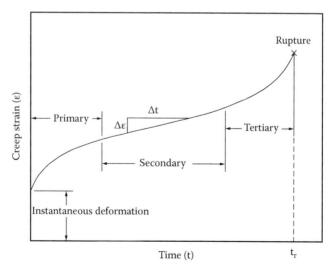

FIGURE 1.12
Typical creep curve obtained in a uniaxial creep test conducted at temperatures above 0.5 T_M (melting temperature).

1.4 Standards for Mechanical Testing

The properties of materials determined from mechanical testing are of interest to many parties ranging from design engineers, metallurgists, producers, and consumers to researchers. Therefore, the mechanical tests must be consistent, uniform, and reproducible in the way they are conducted and interpreted. This is accomplished by following "standards," which define testing machines, specimen shapes and sizes, experimental procedures, and data interpretations. The standards for each of the mechanical test techniques have evolved over the years through coordinated efforts of various professional societies. Some of the popular standards are those of American Society for Testing and Materials, United States (ASTM), International Organization for Standardization (ISO), British Standards (BS), Deutsches Institut für Normung, Germany (DIN), Japanese Industrial Standards (JIS), Bureau of Indian Standards, India (BIS), etc.

It may be noted that one of the earliest British standards, BS18 (1904) of tensile testing of metallic materials, was first published in 1904. The first American tensile testing standard was issued in 1924 (ASTM E8-24T 1924)—subsequently, ASTM E 21 as a high-temperature tensile testing standard in 1931. In Germany, DIN was founded in 1917 and the first tensile test standard was released as DIN 1604 in 1924. With advancements in the methodology of testing and data acquisition (manual to automated and computer controlled), both ISO and ASTM standards are being periodically reviewed and revised through intercomparison exercises or round robin test programs.

1.5 Specimen Dimensions

1.5.1 Tensile, Fatigue, and Creep Test Specimens

Specimen designs for tensile, fatigue, and creep properties have been developed, in general, for the evaluation of realistic and consistent material properties of bulk homogeneous materials by considering a sufficiently large volume of material. The specimen sizes are basically designed to ensure that the sample tested contains sufficient microstructural features (grains, grain boundaries, second phase particles, inclusions, pearlite or lath colonies, dislocation networks, etc.) such that the mechanisms of deformation and failure are influenced by the aggregate of these microstructural features ensuring multicrystal behavior and the properties derived are representative of the bulk material.

In a typical specimen design for tensile testing, shown in Figure 1.13, the specimen gauge diameter is at least 4–8 mm in diameter and a nominal gauge

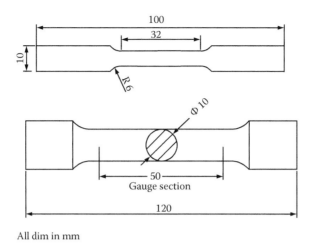

All dim in mm

FIGURE 1.13
Typical geometry and dimensions of flat and round types of tensile test specimens.

length is about 25–50 mm; sizes larger than this could also be used. The gauge length is the region in which deformation and failure take place under the action of tensile stress. The end portions are of higher diameter than the gauge section to allow for gripping without any stress/strain, and a section of smooth radii connects the gripping portion to the gauge section to avoid any stress concentrations In addition to these considerations, to ensure comparison of the elongation of different size specimens, the specimens must be geometrically similar and obey Barba's law (1880). This is satisfied by ensuring that the gauge length is five times the gauge diameter (i.e., 5D) for cylindrical specimens and $5.65\sqrt{Area}$ for flat rectangular specimens.

Specimens for fatigue tests are designed according to mode of loading, which may be axial stressing, rotating beam, alternate torsion, or combined loading. The specimen sizes and volumes in conventional fatigue are very similar to the tensile specimen configuration, with similar conditions for the radius of the fillet (eight times the gauge diameter) and gauge length of at least three to four times the gauge diameter (Figure 1.14a). A minimum cross-sectional diameter of 5 mm is recommended for fatigue specimens. For specimens tested in tension–compression, hourglass types of specimens (Figure 1.14b) are employed; this has the advantage of minimizing buckling in push–pull loading. Specimens for uniaxial creep tests are quite similar in shape and dimensions of tensile test specimens. The gauge length (GL) and diameter of the GL are typically 30–50 and 6–10 mm, respectively.

1.5.2 Fracture Toughness Specimens

The Charpy specimens for impact loading and fracture toughness specimen designs such as the compact tension (CT) specimen have a size criterion

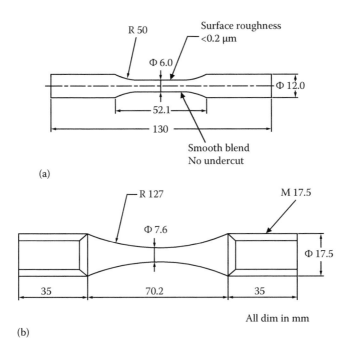

FIGURE 1.14
Typical geometry and dimensions of fatigue test specimens. (a) Round and (b) hourglass types.

for a valid test. Standards specify a Charpy impact specimen of size 10 mm square bars of 55 mm length with a notch machined at the center of one edge and loading under stipulated parameters of specimen mounting and impact velocity of pendulum (Figure 1.10). The energy absorbed in fracturing a notched specimen is a complex function of both the elastic and plastic deformation during crack initiation and propagation from the notch root. These processes are sensitive to stress state and hence this specimen size is very specific and accepted worldwide for determination of the ductile–brittle transition curve.

Standard fracture toughness tests are designed to allow reproducible determination of the relevant fracture characterizing parameter, such as plane strain fracture toughness, K_{IC}, the J-integral, or CTOD. Certain conditions regarding specimen and crack geometry, loading parameters, and shape of load-displacement curve have to be met in the tests before a valid result can be reported. In establishing the specimen size requirements for plane-strain K_{IC} tests, the specimen dimensions should be large enough compared with the plastic zone so that effects of plastic zone are minimal and LEFM is obeyed. The pertinent dimensions of CT specimens for K_{IC} testing (Figure 1.15) are crack length (a), thickness (B), and the remaining uncracked ligament length (W − a, where W is the overall specimen depth). The following

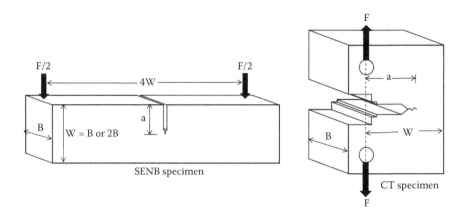

FIGURE 1.15
Single edge notched bend (SENB) specimen and compact tension (CT) specimen geometries for fracture toughness evaluation.

size criteria for a, B, and W are specified for a CT specimen to ensure linear elastic behavior and ensure plain strain conditions at crack tip:

$$a, B, (W - a) > 2.5 \left(\frac{K_{IC}}{\sigma_{ys}} \right)^2 \qquad (1.1)$$

where σ_{ys} is the yield strength of the material. Thus, the specimens for K_{IC} determination must be sufficiently thick enough to establish plain strain conditions. J_{IC} is the measure of the fracture toughness at the onset of slow stable crack extension under nonlinear elastic–plastic conditions. The J integral procedure testing requires samples thinner than K_{IC} specimens by about a factor of 20. Similar dimensional requirements are specified by various standards for a three-point specimen for fracture toughness determination.

It can be summarized that the mechanical tests for evaluating properties such as yield stress, flow stress, ductility, creep, and fatigue properties require specimens typically of 5–8 mm diameter and about 25–50 mm in length per the standards worldwide. For fracture toughness properties, standards demand certain specimen geometry and size criteria resulting in thicknesses above 12–25 mm and other dimensions proportionally larger.

1.6 Need for Specimen Miniaturization

The data from standard mechanical tests have been the mainstay of the codes for the design and manufacture of engineering structures. However, meeting the size requirements for mechanical testing may not be practically possible in situations where only a limited volume of material is available for examination. Such situations may arise in the following cases.

1.6.1 Alloy Design and Development

The design of advanced high-strength and damage-tolerant metallic alloys for applications related to energy, transport, health, and infrastructure applications forms the engineering and manufacturing backbone of our modern society. Examples are materials for radiation-resistant steels for nuclear and fusion power plants, creep-resistant steels, and nickel alloys in power plants and plane turbines; ultrahigh-strength Fe–Mn steels, Ti- and Mg-alloys for lightweight mobility and aerospace design; metallic glasses for low-loss electrical motors and wind energy generators; or biomedical CrCoMo and Ti-implant alloys for aging societies. In addition, composite materials constitute a significant proportion of the engineered materials ranging from everyday products to sophisticated niche applications. The design of such novel engineering alloys is a two-step process, which involves understanding the relations between processing and microstructure evolution, on the one hand, and the relations between microstructures and mechanical properties on the other. In this process, there is always a need for rapid measurement of mechanical properties to quickly sort out the various heats and narrow down from a large range of alloy designs to the desirable choice in the shortest possible time. In most situations, the amount of exotic material prepared on a lab scale may be very limited for preparing standard size specimens or it may be time consuming and economically not feasible to prepare standard specimens from all the heats of the alloy matrix. It becomes compelling to employ mechanical testing using small volumes of available material for such material development programs.

1.6.2 Periodic Assessment and Life Extension of Engineering Components

With rapid industrialization in the last few decades, many of the structural components in power generation, aerospace, chemical, and process plants are nearing their designed lifetime. The material properties of in-service structural components are altered due to changes in microstructure caused by operating stresses, thermal aging, irradiation, and other adverse conditions. This can possibly lead to reduced operational life and premature failure of the component during service. For reliable integrity assessment of structural materials and accurate residual life predictions of the plant component, the knowledge of the changes in their mechanical properties is very essential. It is necessary to obtain data from the component itself so that both metallurgical conditions and residual mechanical properties can be directly assessed in their current state of damage. The evaluation of mechanical properties of in-service components through conventional techniques, such as tensile, Charpy impact, and fracture toughness tests, is destructive in nature as it requires extracting bulky samples from the component and thereby causing considerable damage and compromising its structural integrity. This can result in forced outage and often require some repair actions to allow further operation.

1.6.3 Weld Joints and Coatings

Welding or joining of metals has become indispensable in the field of manufacturing and construction of equipment, machinery, and products. In a welded structure, the material is inhomogeneous and the mechanical properties vary as a function of the distance from the weld bead. Apart from the weld and base metal, there exists a very narrow region on the base metal, referred to as the heat-affected zone (HAZ), whose properties are influenced by the welding process and the postweld heat treatments. Assessing the mechanical behavior of the individual HAZs, which are typically about 0.5–1.0 mm in width, is not possible using standard specimen sizes as the volume of the homogeneous material available is very limited.

Similar limitations of insufficient material for mechanical property evaluation also exist in coating and hard-faced components. The coating and the hard facing are forms of surface treatment where the bulk material surface is given a protective layer of another material having more superior properties than those of the bulk material.

1.6.4 Failure Analysis

Failure analysis and prevention are key aspects of all engineering disciplines. The materials engineer often plays a lead role in the analysis of failures to ascertain whether a component or product failed in service or if failure occurred in manufacturing or during production processing. Failure analysis is a multidisciplinary subject involving mechanics, physics, metallurgy, chemistry, manufacturing processes, stress analysis, design analysis, and fracture mechanics. The investigations mainly pertain to (1) surface of fracture, (2) geometry and design, (3) manufacturing route and processing, (4) adequacy of physical and mechanical properties of material, (5) applied loads and residual stresses, (6) service conditions, and (7) environmental conditions.

In addition to various nondestructive inspections (such as X-ray radiography, magnetic particle inspection, and fluorescent penetrant, ultrasonic, and eddy current inspections) and metallographic examinations, mechanical testing is often a requirement in the process of investigating the failure. The decision to remove a sample specimen for chemical, microscopic, and mechanical testing is a very contentious issue in failure analysis investigations. Samples selected should be characteristic of the material and contain a representation of the failure or damage. Also, a sample is taken from a sound and normal section for comparative purposes. Barring hardness measurement, estimation of other mechanical properties tensile strength, elongation, fracture toughness, etc., is rarely done as it requires large volumes, which are seldom available in a failed component.

1.6.5 Micro- and Nanodevices

The ongoing miniaturizations in many areas of modern technologies such as microelectromechanical systems (MEMS), nanoelectromechanical systems (NEMS), and biomaterials has increased the attention toward mechanical properties in the micro- and nanometer regime. Nanocrystalline metals and alloys with average and range of grain sizes typically smaller than 100 nm have been the subject of considerable research in recent years because of their outstanding mechanical properties of superstrength, high ductility, and tribological performance. Such materials are now finding applications in a wide range of fields, such as the aviation, automotive, and electronics industries. MEMS devices use materials such as silicon and other thin films that have not been completely characterized with respect to mechanical properties. Thus, the evaluation of mechanical properties of nanostructured materials has become necessary for design of MEMS devices and for commercial exploitation of the MEMS technology.

1.7 Miniaturized Specimen Testing—A Genesis

The miniature (or small) specimen mechanical test technology refers to methods of characterizing the mechanical behavior of materials using specimen sizes much smaller than the standard specimen sizes. The development of mechanical test techniques using small volumes of material has been a research topic for a little more than three decades. It will be interesting to see that this technology first attracted attention in the nuclear industry during the materials development programs for nuclear fission and fusion reactors in the late 1970s. The following section attempts to capture the genesis of this technology.

The development of small-specimen testing technology is closely related to the growth of nuclear power technology. Through the 1960s and 1970s, many countries embarked on programs to build nuclear reactors for the production of electricity. Nuclear reactors use the heat from fission of uranium and plutonium isotopes to produce steam, which turns turbines to generate electricity. The components common to all nuclear reactors include fuel rods, assembly, control rods, a coolant, a pressure vessel, a containment structure, and an external cooling facility. It is the neutron energy sustaining the chain reaction that classifies the reactor as thermal or fast type. While most of the first-generation reactors were based on water-cooled thermal neutron-based technology, the development of fast neutron and breeder reactors also gained prominence due its capability to extract maximum amounts of energy from nuclear fuel.

The core of a nuclear power reactor is where the fuel is located and nuclear fission reactions take place. The materials used to encase the fuel in fuel rods, to hold fuel rods together in fuel assemblies, and to hold fuel assemblies in place are all considered as core structural materials and so are the materials used in core supporting structures such as pressure vessels, control rods, and core monitoring instruments. The materials in and around the core are in general subjected to a very adverse environment of neutron flux (typical values: 10^{15}–10^{17}n/cm^2/s), stress, and temperatures ranging from 300°C to 800°C depending on the type of the fission reactor. A measure of the effect of irradiation on materials is the number of times an atom is displaced from its normal lattice site by atomic collision processes. This is quantified as displacements per atom (dpa). The microstructural defects produced as a result of displacement damage lead to observable changes in physical and mechanical properties such as hardening, embrittlement, radiation-induced growth and swelling, creep, and phase transformations.

In a typical light water reactor, zirconium alloys are the most commonly used material for fuel cladding and assembly structure, while the reactor core is bounded by a heavy-section steel reactor pressure vessel (RPV) to safely contain coolant water at temperatures of little less than 300°C at pressures ranging from 5 to 15 MPa. During the lifetime of these reactors, which spans over 40–60 years, the RPV internal steel accumulates relatively high, fast neutron fluence. Irradiation by neutrons changes the properties of the materials (e.g., the ductility and fracture toughness of the material). For assessment of the structural integrity and remaining lifetime of RPV internals, as well as for life extension purposes, the need for relevant and valid data on the properties such as fracture toughness of steels and their evolution with neutron fluence was felt. For such an assessment, surveillance programs comprising irradiating mechanical test coupons of pressure vessel steel alongside the fuel in the core of the reactor, were initiated.

In the case of fast fission reactors, the design parameters for achieving high burnups (~150 GWd/t as compared to 20–40 GWd/t in thermal reactors) places severe performance demands on materials used in reactor fuels, reactor core components, and reactor vessels. The reactor core components made of austenitic stainless steels and ferritic-martensitic (FM) steels are subjected to temperatures of about 450°C–700°C and experience a neutron damage of more than 100 dpa against a few tens of dpa in thermal reactor core components. In such extremely demanding operating conditions, both austenitic and FM steels have limitations—the former due to swelling and the latter with respect to elevated temperature strength. Thus, the development of irradiation-resistant improved versions of austenitic and FM steels as well as exotic materials for core components of fast reactors was necessary for pushing the fuel burnup to ~200 GWd/t and beyond so as to improve the economics of power from fast reactors.

Toward the late 70s, the scientific community was also engaged in the efforts to harness the fusion energy in a controlled manner for the benefit of mankind. Fusion of light elements such as deuterium and tritium (D-T), the hydrogen isotopes, in plasma under magnetic confinement (Tokamak) results in virtually inexhaustible energy. In a typical conceptual fusion reactor, the neutrons generated from the D-T fusion reaction are absorbed in a lithium blanket surrounding the core, which transforms into tritium and helium. The kinetic energy of the high-energy (14 MeV) neutrons is absorbed by the blanket and the heat energy is extracted by the coolant (water, helium, or Li-Pb eutectic) flowing through the blanket, from which electricity is generated by conventional methods. In fusion reactors, the plasma facing (first wall, divertor) and breeding-blanket components are exposed to plasma particles and undergo damage due to 14 MeV neutrons. These energies are very high compared to 1–2 MeV in the fission neutron spectrum and the gas produced by transmutation nuclear reactions is also higher ~75 a ppm compared to a few parts per million in fission reactors. Materials to be used in fusion reactors need to operate at higher temperatures (500°C–1000°C) and experience damage of ~30–100 dpa. Development of radiation-resistant materials to withstand the damage induced by high-energy (14 MeV) neutrons became an integral part of the efforts to support the fusion program. The essential part of the material development program was irradiation material testing of candidate alloys focused toward fundamental understanding of the effects of high-energy neutron irradiation on the microstructure and mechanical properties leading to the development of a comprehensive database of irradiated material properties.

Irradiation testing of materials thus became a very crucial aspect for both surveillance of light water reactor structures and material development programs of fast reactor and fusion reactor systems (Lucas 1990). Successful materials development required irradiation testing facilities that reproduce as closely as possible the thermal and neutronic environment expected in a power-producing reactor. However, practical difficulties were soon encountered in such irradiation programs due to limited irradiation volumes available in the commercial reactors or in test reactors. For example, the relatively large size of a full-scale Charpy V-notch (CVN) impact specimen (typical dimensions: 10 mm × 10 mm × 55 mm) restricted the number of specimens that could be irradiated in a given volume under conditions of uniform temperature and neutron flux. In some cases, the archive material of the pressure vessel steel available with the manufacturer was insufficient for a large-scale test irradiation. The space limitations for accommodating conventional mechanical test specimens were more severe, especially in the fast reactors, which are very compact and tightly packed. Irradiation experiments of such large size specimens in the high-flux regions were often very costly as it would require many years of exposure to simulate the damage levels typical of fast reactor service applications.

With prototype fusion reactors seldom available for irradiation material testing, the irradiation behavior data were sought from fission reactors, especially the fast reactors and accelerator-based high-energy neutron source facilities. However, the extremely limited irradiation volume with uniform temperatures and neutron flux necessitated the miniaturization of the specimens that could be fit into the available volume for irradiation. The reduction in specimen sizes was essential to cater to both material development and surveillance programs. Thus, it can be categorically stated that small specimen testing technology has primarily evolved out of the needs of the nuclear power community to develop and monitor the materials in fission and fusion reactor systems.

1.8 Spin-Off Applications of Specimen Miniaturization

The specimen miniaturization has led to several spin-off applications in both the nuclear and non-nuclear industries (Hyde et al. 2007; Lord et al. 2010). One immediate fallout of this idea was the significant reduction in the activated gamma radiation from the small specimens (as compared to gamma dose from conventional size specimens), resulting in considerable ease of handling the material during postirradiation examination. The postirradiation examinations are normally carried out in shielded facilities, called hot cells or lead cells, using remote handling tools. The specimen miniaturization permits relatively easy specimen handling with reduced shielding requirements and therefore reduced cost. Because the neutron damage and irradiation temperature parameters are more uniform across the specimen volume, the measured data from small specimens become a meaningful representative of the corresponding irradiation conditions.

The concept of small specimens for evaluating mechanical properties is very appealing for residual life/structural integrity assessment of aging components in power plant and chemical process industries. This is because it enables the evaluation of mechanical properties of in-service structural components using small volumes of scooped out specimens in a minimally invasive manner so that the component can be maintained in service. This technology now has applications in many other areas where conventional mechanical test methods are practically difficult (as discussed in Section 1.6), such as in new material development, failure analysis, study of weld joints/coatings, micro/nanodevices, etc.

1.9 Concluding Remarks

In this chapter, an attempt has been made to provide the reader a flavor of various conventional mechanical test methods in practice and the requirements of specimen sizes as per the standards. This was followed by highlighting the different situations where size requirements as per the standards cannot practically be met and the need for testing with specimen sizes smaller than the conventional/standard designs. The focus of the next two chapters will be on characterization methods that have evolved over the years for tensile, impact, and fracture toughness properties using specimen sizes of typically 0.3–5.0 mm. The technological relevance of small specimen testing for aging assessment, advanced materials and testing of narrow zones, miniaturized components, etc. is presented in the last two chapters.

References

ASTM E8-24T. 1924. Standard test methods for tension testing of metallic materials.

Barba, M. J. 1880. *Mem. Soc. Ing. Civils,* Part I, p. 682.

BS18. 1904. Forms of standard tensile test pieces. Published by the Engineering Standards Committee, London.

Callister, W., Jr., and Rethwisch, D. G. 2011. *Fundamentals of materials science and engineering—An integrated approach.* New York: John Wiley & Sons Inc.

Dieter, G. E., Jr. 1990. *Mechanical metallurgy.* New York: McGraw–Hill.

Hyde, T. H., Sun, W., and Williams, J. A. 2007. Requirements for and use of miniature test specimens to provide mechanical and creep properties of materials: A review. *International Materials Review* 52 (4): 213–255.

Kuhn, H., and Medlin, D., eds. 2008. *ASM handbook: Mechanical testing and evaluation,* vol. 8. Materials Park, OH: ASM.

Lord, J. D., Roebuck, B., Morrell, R., and Lube, T. 2010. Aspects of strain and strength measurement in miniaturized testing for engineering metals and ceramics. *Materials Science Technology* 26: 127–148.

Lucas, G. E. 1990. Review of small specimen test techniques for irradiation testing. *Metallurgical Transactions A* 21A: 1105–1119.

2

Miniature Specimen Testing for Tensile and Plastic Flow Properties

2.1 Introduction

In the previous chapter, it was seen that the idea of using small specimens for mechanical testing had actually originated in the nuclear industry to cater to the irradiation material testing programs. Small or miniaturized specimens could be fit in the available space in nuclear reactors and also could be extracted from an irradiated component and handled with much reduced radiological hazard for examination. As a spin-off, this technology soon found applications in many other areas where conventional mechanical test techniques were practically difficult, such as in failure analysis, study of weld joints and coatings, assessment of residual life of power plant components, and in development of new alloys and exotic materials where limited volume of material is available.

A number of different miniature specimen test techniques that have evolved over a period of time can be broadly classified into two categories: (1) those based on direct scaling down of the conventional specimen geometry, and (2) those based on unconventional specimen geometries and novel loading techniques. The first category includes tests employing specimen similar in shape and form of full-scale tensile, fracture toughness, fatigue, and impact specimen but whose dimensions are proportionately reduced (also referred to as subsize specimen). The loading configuration in this category of tests remains the same as that of conventional tests.

The other class of miniature specimen tests are novel techniques where specimen geometry and loading configurations are unconventional and do not have large specimen conventional equivalents. The simplest specimen configuration in this category is the disk specimen with sizes ranging from 3.0 to 8.0 mm in diameter and thicknesses of 0.3 to 1.0 mm. The most popular tests using disk specimens are (1) punch tests (shear punch and small punch test) and (2) spherical/ball indentation tests. These test techniques involve loading a small disk specimen with a specifically shaped indenter and analyzing the resulting load-displacement data obtained during the deformation process for determination of plastic flow and fracture properties. This chapter is devoted to small specimen test methods for determination of tensile and associative flow curve properties and will focus on the developments in subsize tensile and other novel test methods such as shear punch, small punch, and ball indentation.

2.2 Tensile Tests with Subsize Specimens

It was seen in the previous chapter that the conventional specimen designs according to the standards for tensile testing are typically of 4–8 mm in diameter or width with a gauge length in the range of 25–50 mm or more. These specimen sizes provide the complete stress–strain, proof stress, ultimate tensile strength (UTS), ductility parameters such as uniform and total elongations, and true stress–true strain curves representative of the bulk material.

Tensile tests with subsize specimens have the potential to provide the stress–strain data representative of bulk material behavior through careful consideration of various influencing factors such as microstructure and its relation to the specimen dimensions, specimen preparation methods, specimen alignment in the load train, resolution/accuracy of load, and strain measuring sensors. A typical subsize tensile specimen of the standard ASTM E8 and a further miniaturized version of the subsize specimen are shown in Figure 2.1.

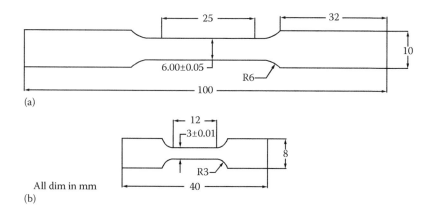

FIGURE 2.1

Schematic of (a) subsize tensile specimen of ASTM E8 and (b) a further miniaturized version of the specimen with gauge length of 12 mm and gauge width of 3 mm.

2.2.1 Influence of Thickness-to-Grain Size Ratio

The flow stress of a material is influenced by the microstructural features such as the grain size, grain shape and orientation, texture, precipitates and their size relative to the test piece dimensions. Grain size, number of grains, and gauge section geometry are important factors to be considered for comparing the data between large and subsize specimens. Most studies have shown that at least 20 grains in the cross section are sufficient to permit accurate prediction of bulk tensile properties by miniaturized specimens (Armstrong 1961). The critical thickness of miniature tensile specimens is expressed in terms of thickness-to-grain size t/d_g ratio. When t/d_g is generally above a certain critical value (of about 10–12), multicrystal response is maintained, enabling good agreement between the large and scaled-down specimens (Miyazaki, Shibata, and Fujita 1979). Typical critical t/d_g ratio reported for stainless steel of types AISI 304 and AISI 316 is around 6 (Igata et al. 1986).

For specimen designs with t/d_g ratio less than critical value, the global flow stress of the material decreases (Figure 2.2) with specimen width and thickness. This has been explained by the long-range interaction model (Miyazaki et al. 1979). In a thicker sample of polycrystalline material, the grains constrain each other to accommodate the misfits between them in the form of accumulated dislocations inducing the long-range back stress fields. In thinner samples with few grains across the thickness, the loss of constraint for the grains close to the surface relaxes the back stresses considerably and causes a reduction in the flow stress. While this is typical for single phase

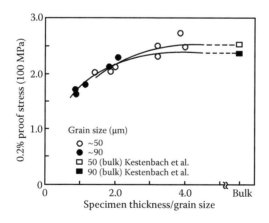

FIGURE 2.2
Plot showing the dependency of 0.2% proof stress on the ratio of specimen thickness to grain size for AISI 316 stainless steel. (From Igata, N. et al., *Journal of Nuclear Materials* 122 & 123: 354–358, 1984.)

metals such as pure Al, Cu, Fe, Al-Cu, and type 304, 316 steels, for steels with tempered martensitic or baintic microstructures, the flow stress is more influenced by the lath size and the strength reductions in thin cross sections ($t < 10$ d_g) are relatively much smaller than the former.

The ductility defined as the percentage change in the gauge length (GL) comprises of two components: uniform elongation and necking elongation. The uniform elongation depends on the metallurgical condition of the material (i.e., strain-hardening exponent), while the necking elongation depends more on specimen shape and size. The necking elongation decreases as the ratio of cross-sectional area to gauge length is reduced and hence total elongation also follows the same trend. To enable comparability of ductility values between different sizes of bulk samples, ASTM E 8 and ISO 6892 recommend proportional geometries (GL = 5.65 $\sqrt{Area}$ for flat specimens and GL = 5D for round samples, where D is the gauge diameter) based on Barba's law of geometric similarity (Dieter 1961). The same criterion needs to be followed for the comparison of ductility data between large and subsize specimens.

The studies of thickness effects on ductility and strain components have been rather limited. In one such study by Byun et al. (1998) on reactor pressure vessel steel (SA 508), thickness effect on ductility was evident for specimen thickness below 0.2 mm. This resulted from the preferred deformation in the thickness direction as compared to the width direction. This preferred deformation was explained based on the stress state within the thin specimen caused by the differences in the magnitude of back stress in the thickness and width directions leading to flow localization in a preferred direction even before the development of plastic instability. The necking angle, defined as the angle between the loading line and the fracture plane was also thickness dependent, decreasing from 90° to 60° as the thickness decreased from 2 to 0.1 mm. The morphology of the fracture surface was also shown to be dependent on specimen thickness, with shallow, elongated dimples in thin specimens and equiaxed dimples in the thick specimen. In general, specimen thickness does not influence strength and ductility when thickness is above 0.2 mm and t/d_g is greater than 10–12. These are general considerations for the subsize tensile test data to represent the bulk material behavior.

2.2.2 Ultra Subsize Specimen Designs

In subsize tensile testing, many variants of the regular sample designs explored, especially with one or more dimensions (width/thickness) of the specimen reduced below 0.5 mm. LaVan and Sharpe (1999), have successfully employed a dog-biscuit-shaped $3.0 \times 0.2 \times 0.2$ mm specimen in the gauge section with V-shaped ends (Figure 2.3) that can be conveniently located in the grips mounted in linear air bearings to reduce the friction of the grips. Combining this with the interferometric strain/displacement gauge, Sharpe,

FIGURE 2.3
A microsample in the grips, with the movable grip supported by air bearing. (From LaVan, A., and Sharpe, W. N., *Experimental Mechanics* 39:210–216, 1999.)

Danley, and LaVan (1998) demonstrated good repeatability of the stress–strain curves of the microtensile sample tests and the measured yield and ultimate tensile strengths were only 5% lower than the macrosample test results for A 533-B steel. The microtensile specimen design employed by (Konopik and Dzugan 2012) has a gauge length of 3.0 mm and gauge width of 1.50 mm (Figure 2.4) machined from an 8.0 mm diameter and 0.5 mm thick disk specimen. The advantage of such specimen designs is that the load-displacement curves have the same shape and features as in the case of standard tensile test, making the interpretation of the data straightforward.

A dumbbell-shaped miniature tensile specimen was designed by Partheepan, Sehgal, and Pandey (2006) from an 8.0 mm diameter disk (Figure 2.5) with a gauge width of 1.0 mm and thickness of 0.5 mm that can be pulled in tension. As the gauge width is nonuniform in the deforming specimen, the interpretation of the load-elongation curve is not straightforward like the tensile test data. A novel identification strategy was employed by the investigators based on the optimal matching of a miniature test load-elongation curve with the

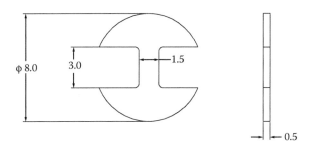

FIGURE 2.4
Microtensile sample investigated by Konopik and Dzugan. All dimensions in millimeters.

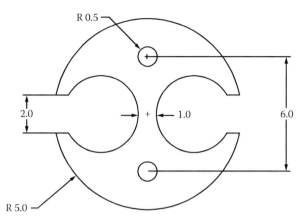

FIGURE 2.5
Configuration of a miniature dumbbell specimen employed by Partheepan et al. All dimensions in millimeters.

finite element predicted load-elongation curve in a piecewise linear manner up to the peak load and thereby predicting the constitutive behavior of the unknown material. The results obtained using the inverse finite-element-based algorithm for strength as well as uniform elongation from dumbbell specimen testing compared well with the conventional tensile test results for various steel alloys.

Ring tensile specimen geometry is a common design for characterizing the deformation behavior of thin-walled tubular components in the transverse (hoop) direction. The photograph of a typical ring tensile specimen (Kim and Kim 2013) is shown in Figure 2.6. The experimental device consists of two semicylindrical beams in contact with the inner surface of the sample, moving relative to each other when the load is applied. Often a centering piece is located inside the ring specimen in order to prevent the reduced region from bending at the beginning of the test. The determination of the

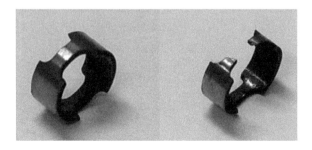

FIGURE 2.6
Ring tensile specimen geometry of a tubular specimen. (From Kim, J. H., and Kim, S. H., *Journal of Nuclear Materials* 443:112–119, 2013.)

stresses usually requires computational analysis and an assumption of the friction between the specimen and grip/mandrels.

2.2.3 Challenges in Subsize Tensile Testing

2.2.3.1 Specimen Machining and Gripping

When specimen dimensions become small, several precautions must be observed in specimen preparation techniques and handling for testing. The fabrication technique chosen should enable good control of specimen dimensions without introducing deformation, surface burrs, or microcracks around the edges. When the volume of affected material arising during machining of subsize specimens becomes a significant percentage of the specimen volume, the damaged layer can significantly influence the flow behavior, leading especially to the early onset of necking. The preferred method of machining subsize tensile specimens is electrodischarge machining (EDM) followed by diamond grinding and polishing to remove the recast layer to achieve good dimensional tolerances. Flat types subsize tensile specimens have been generally preferred due to easy machining from available stock and handling during testing as compared to the round or wire specimens.

Grips for holding the specimens also need to be miniaturized and customized, consistent with the specimen dimensions. Small misalignments create large bending strains, and gripping can easily alter or destroy the sample. Figure 2.7 is a photograph of customized wedge action grips for

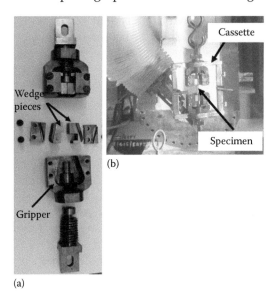

FIGURE 2.7
Photograph of (a) wedge action grip for holding subsize tensile specimen and (b) cassette holder for handling the gripped specimen.

elevated temperature testing of a neutron-irradiated subsize tensile specimen remotely in a hot cell (Figure 2.1b) and a cassette holder for safe handling and loading of the gripped sample onto the test machine.

2.2.3.2 Strain Measurements

As the loads and deflections are small in subsize specimen tensile testing, a well aligned precision test frame is essential. With smaller gauge lengths (<10 mm) and thinner cross sections, fitting conventional extensometers to the specimen often becomes practically impossible. The specimen elongation is often approximated by cross head displacement and must be appropriately corrected for compliance of the machine frame and grips for meaningful data analysis.

For accurate and reliable strain measurements during subsize tensile testing, noncontact techniques such as laser-based extensometer/interferometry and digital image correlation (DIC) have gained popularity. A summary of the various noncontact methods for strain measurement during miniaturized tensile testing is presented by Lord et al. (2010). Digital image correlation is a full-field image analysis method, based on gray value digital images, which can determine the contour and the displacements of an object under load in three dimensions. In the DIC method (Figure 2.8), the image of the random pattern applied on the test piece is acquired by a digital camera as the deformation progresses. By comparing the images and by tracking the blocks of pixel at different instants of deformation, the surface displacement is measured. The full-field two-dimensional (2D) and three-dimensional (3D) deformation vector fields and strain maps are calculated using sophisticated correlation functions from the position of the center of the pixel blocks. DIC has the advantage of full-field capability and measures the strain well beyond the necking, capturing the true stress–true strain behavior beyond the ultimate tensile stress.

The interferometric strain/displacement gauge (ISDG) technique is another noncontact method for strain measurements in subsize tensile testing where the relative distances between two gauge marks are measured optically. The method involves making marks (like tiny pyramidally shaped indentations) that, when illuminated with laser, produce fringe patterns caused by interference of diffracted reflections from the facets of the indentations. With specimen displacement, the fringe patterns also move and are converted into electrical signals and related to strain. Employing this technique, Sharpe et al. (1998) achieved a typical resolution of ~5 microstrain for 0.3 mm gauge length steel samples and demonstrated stress–strain curves similar to those found in standard methods.

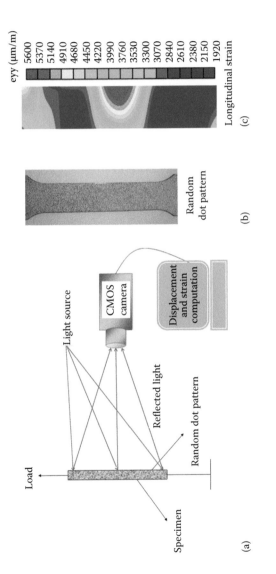

FIGURE 2.8

Schematic of the (a) digital image correlation technique for strain measurement; (b) speckle pattern on the specimen; and (c) strain distribution.

2.2.4 Micro- and Nanoscale Tensile Testing

The previous sections clearly indicated that the mechanical response of materials varies with length scale. When specimen dimensions are sufficiently larger than microstructural scale (grain size, defects), laws of continuum are obeyed and the deformation is representative of the bulk material. However, materials in miniaturized devices such as microelectromechanical system (MEMS) are used in the form of thin films, the thicknesses of which often remains comparable to the microstructural length scale.

The synthesis of materials with reduced dimensions typically less than 100 nm, such as coatings, thin films, and nanocomposites, which form mechanical structures of MEMS, is an intense area of research because of their appealing properties, such as high strength, enhanced tribological and environment resistance, and applications ranging from biomimetics to aerospace. It becomes imperative to evaluate the mechanical response of materials at length scales that are similar to those used in actual application. In such situations where there is an overlap between the length scale of defects controlling the deformation of materials with that of the specimen dimensions, the laws of continuum mechanics are not obeyed and mechanical properties of materials evaluated in small volumes differ significantly from those evaluated in bulk form due to both geometric and microstructural constraints. Such studies are important not only from the point of qualifying the device performance for reliability but also for unraveling the deformation mechanism of materials when atomic scales are approached.

Juxtaposed with processing and synthesis, testing at the micro- and nanoscale (Gianola and Eberl 2009) poses many technical challenges with respect to positioning and handling of the specimen, accurate application of force, and reliable measurement of load/displacement during deformation. In addition to tensile tests, mechanical testing at micro- and nanoscales has diversified over the last decade—for example, bending/curvature, pillar compression, depth-sensing indentation, atomic force microscopy (AFM)-based cantilever methods. A variety of advanced microfabrication methods (Jaya and Alam 2013), such as focused ion beam (FIB) machining, computerized numerically controlled (CNC) machining, electrodischarge machining (EDM), laser-based processes, lithography, lithographie galvanoformung abformung (LIGA), and electrodeposition have been adopted for fabrication of specimens at micrometer and submicrometer length scales. While photolithographic and LIGA techniques involve building the system layer by layer using a mold, micro-EDM, laser cutting, reactive ion etching, electropolishing, and FIB involve machining or removing material from the bulk system to the required shape.

Two approaches have evolved for handling micro- and nanoscale specimens that cannot withstand the brutality of conventional handling tools. The first method involves integrating a support structure into a microfabrication scheme to anchor the thin film to a substrate that can be handled easily with tweezers for integrating into the tensile loading platform. By removing the support structure in the gauge section, it is ensured that the loads are incurred by the thin film only. The other approach involves cofabrication of the specimen and testing apparatus, as demonstrated by Haque and Saif (2002). They employed a microfabricated tensile test chip where the freestanding thin film specimen (30–50 nm thick) is cofabricated with a force sensor beam made of single crystal silicon. A piezoactuator pulls one end of the chip deforming the specimen whose other end is attached to a supporting beam assembly. Cofabrication has advantages of circumventing gripping and alignment issues by incorporating the tensile loading into device fabrication apparatus.

When the largest specimen dimensions get down to a few tens of micrometers as in the case of nanotubes and nanowires, in situ microscopic techniques (Figure 2.9) that make use of imaging and diffraction are employed for direct observation of the deformation. Demczyk et al. (2002) employed microfabricated tensile testing for direct application of a tensile strain to individual carbon nanotubes while they viewed the defect structures and lattice deformation in situ in a transmission electron microscope (TEM). By integrating a piezoelectric drive connected to a linear spring into a TEM, they observed the actual breakage of a nanotube in tension.

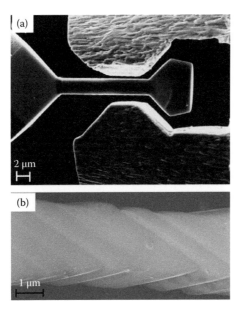

FIGURE 2.9
(a) SEM image of a single-crystal copper tension sample aligned in a tungsten sample gripper. (b) High-resolution SEM image of the tension sample with distinct glide steps visible on the sample surface. (From Kiener, D. et al., *Acta Materialia* 56:580–592, 2008.)

2.2.4.1 Actuation and Force/Displacement Measurement

A conventional macrotensile testing machine consists of a test frame housing a load train, a hydraulic or a screw-driven actuator for straining the sample, and a data recording system. Commercial load cells with a load resolution of 0.001 g and displacement sensors (strain gauge, linear variable differential transformer [LVDT], capacitance, etc.) with typical resolution of 0.1 μm are commonly employed in macrotensile testing. However, the instrumentation involved in a microtensile test machine is significantly more sophisticated than in its macrocounterpart.

The major limitation of nanometer and subnanometer resolution is due to the stick–slip effect resulting from static friction, a finite system stiffness, and play in bearings due to manufacturing tolerances. A method resolving the minimal necessary step size is the use of the so-called inertial drive combined with piezo-electric actuators (Zhou et al. 2006). As regards the measurement of tiny loads, conventional load sensors based on resistive, capacitance, or inductive techniques can measure forces as small as a few tens of micronewtons. AFM-based techniques have enabled measurement of forces below micro-/nanonewton range. In the AFM, the sample is scanned by a tip, which is mounted to a cantilever spring and the force between the tip and the sample is measured by monitoring the deflection of the cantilever (Butt, Cappella, and Kappl 2005).

It can be summarized that tensile testing at small size scales is an attractive area of research and application as the results can be directly interpreted similarly to conventional tensile test procedures. The results obtained through subsize specimens can be scaled to predict bulk material behavior as long as the smallest dimension of the specimen (e.g., thickness) is above 0.2 mm and contains at least 10–12 grains. In the area of micro- and nano-tensile testing, rapid progress has been made in advanced experimental techniques enabling instrumented testing in a very controlled manner for understanding the deformation mechanism at atomic scales.

The following sections of this chapter will deal with nonconventional test techniques for probing the plastic flow behavior of materials.

2.3 Shear Punch Test

The shear punch test was first developed by Lucas (1983) and his co-workers to extract both strength and ductility from thin sheet samples as small as TEM disk specimens (3.0 mm diameter and 0.3 mm thick). This technique is based on a blanking process common to sheet metal forming. The test involves blanking a small disk specimen clamped between a set of dies using a flat ended cylindrical punch as shown schematically in Figure 2.10. The key dimensions in the shear punch setup are the punch diameter (D_{punch}), specimen thickness (t), die diameter (D_{die}), and the clearance zone width (c).

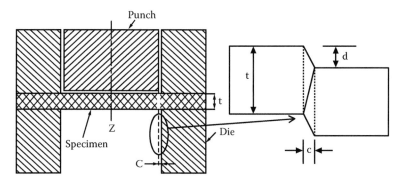

FIGURE 2.10
Schematic of shear punch test technique.

Various combinations of specimen thickness and punch diameter have been adopted by different researchers with D_{punch} ranging from 1.0 to 6.25 mm, t in the range of 0.2–0.75 mm, and the nominal punch-die clearance width in the range of 0.015–0.030 mm, which is about 5%–10% of specimen thickness.

During the slow blanking operation, the load is recorded as a function of the punch travel and the resulting load-displacement curve (LDC) is analyzed to evaluate the strength and ductility parameters. A typical load-displacement curve, shown in Figure 2.11, has features similar to the conventional tensile test—namely, a linear region, followed by nonlinear increase of

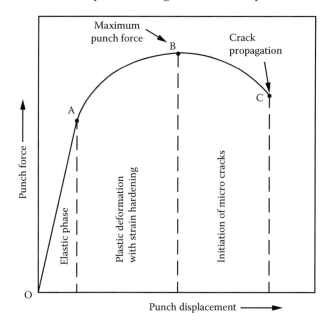

FIGURE 2.11
Typical load-displacement plot obtained in shear punch test.

load with displacement up to maximum load and decrease in load leading to failure. This enables direct comparison and correlation of the shear punch test results with that of a conventional tensile test (Lucas et al. 1984).

The deformation regimes in shear punch (ShP) tests have been systematically analyzed by Lucas, Sheckherd, and Odette (1986) by sectioning a brass sample at various points along the LDC and viewing the etched grain structure. It is seen that the deviation from linearity of the initial linear portion of the curve corresponded to the onset of permanent indentation of the specimen by the punch. Beyond this region, as the punch penetrates, the material work hardens to compensate for decrease in load-bearing thickness and the load increases nonlinearly with punch displacement. A number of slip line fields calculated to analyze the flow in the process zone indicate shear as the principal stress state with contributions from tension and bending. At the point of plastic instability where the strain hardening does not compensate for specimen thinning, load falls with increasing displacement, leading to the separation of the punched piece. For ductile materials, fractographic analysis of the punched disk at the onset of plastic instability has indicated the nucleation of microvoids and their coalescence to form macrocracks indicative of a final shear type of ligament failure.

2.3.1 Experimental Methods

The experimental test setup and the method of analyzing the LDC have evolved over the years. The test fixture essentially consists of a set of dies between which the specimen is clamped. A shear-punch test is generally performed by placing the test fixture between the compression platens of a universal testing machine (UTM) and forcing the flat cylindrical punch through the specimen at constant speed (typically 0.1–1.0 mm/min).

In the initial period of its development, investigators used the cross-head movement to approximate the specimen displacement. However, it was found that the elastic deflections or compliance of the test frame and fixturing had a profound influence on the initial slope of the LDC, often posing difficulty in interpreting the onset of yielding. The influence of the machine compliance on the experimental data was minimized to a certain extent by measuring the punch displacement using either an extensometer across the test fixture (Karthik et al. 2002) or a displacement sensor connected to the moving punch (Toloczko, Kurtz et al. 2002). Later, Karthik et al. (2009) shifted the point of displacement measurement from moving punch to the specimen bottom (Figure 2.12) and demonstrated that the effect of punch and die compliance on the resulting LDC could be completely eliminated in this modified experimental setup.

2.3.2 Analysis of Load-Displacement Curve

As the LDC of a shear punch test (Figure 2.11) is similar to that of a conventional tensile test, the representative strength properties (yield and maximum) are obtained by analyzing the LDC using methods similar to tensile test analysis.

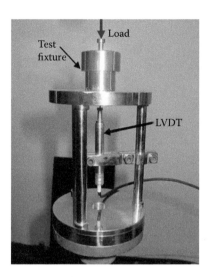

FIGURE 2.12
Shear punch experimental setup with LVDT connected to specimen bottom for displacement measurement.

The method of determining the yield stress in a shear punch (ShP) test and its correlation with tensile yield strength (YS) has been a subject of research and debate over the years. The point of deviation from linearity of the initial portion of LDC was first used as an approximate measure of the shear yield load (Lucas et al. 1986). However, for materials exhibiting a very smooth transition from the linear to the nonlinear deformation, this method of locating the yield load resulted in considerable scatter. Online acoustic emission (AE) monitoring was employed by Kasiviswanathan et al. (1998) in ShP test procedures for accurate prediction of the yield load. Acoustic emission occurs during material deformation due to rapid release of transient energy from localized sources such as regions of stress and strain fields. The online AE signal generated during specimen yielding in shear punch aided in accurate identification of shear yield load and was found to reduce the error associated with shear yield stress determination. Toloczko, Abe et al. (2002) proposed the method of measuring yield stress at an offset shear strain analogous to the offset procedure used in tensile testing.

The finite element method (FEM) has been one of the powerful tools for understanding the deformation response of small specimen tests like shear punch where stress states are nonuniform across the specimen volume. Using finite element (FE) simulation and analysis, the onset of yielding in shear punch was rationalized by Guduru et al. (2006) based on the spread of the plastic zone across the specimen thickness. Specimen yielding has been shown to occur at a shear stress corresponding to 0.15% offset similar to 0.2% offset yield stress in tensile test. However, a larger percentage offset (1% offset) was required to achieve the same yield stress in actual experiments as

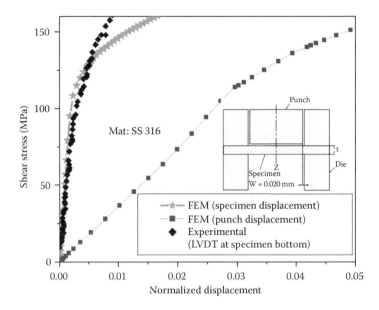

FIGURE 2.13

Comparison of the elastic portion of the load-displacement plots of shear punch experiments with that generated from finite element analysis.

the FE-generated curve was much steeper than the experimental curve due to the finite compliances of the die and punch components. In the experimental setup of Karthik et al. (2009) where the point of displacement measurement was at the bottom of the specimen, the experimental curves overlapped well with the FE curve (Figure 2.13), thus validating the 0.15% offset definition for yield stress (Figure 2.14).

Most of the numerical studies of shear punch using FEM have been limited to yield stress, due to complexities of extreme magnitude of the deformations in the narrow clearance zone beyond yield stress. The extremely large, localized deformations cause excessive distortion of the elements in the mesh, degrading the accuracy of the finite element approximation and require robust remeshing procedures.

As compared to the identification of the yield load in shear punch, the maximum load, P_u, is easily discernible from the point of peak of the load-displacement curve. The representative shear punch strengths, τ, corresponding to the yield and maximum load are estimated using the following equation:

$$\tau = \frac{P}{2\pi r t} \tag{2.1}$$

where P is the yield load or maximum load, t the specimen thickness, and r the punch radius.

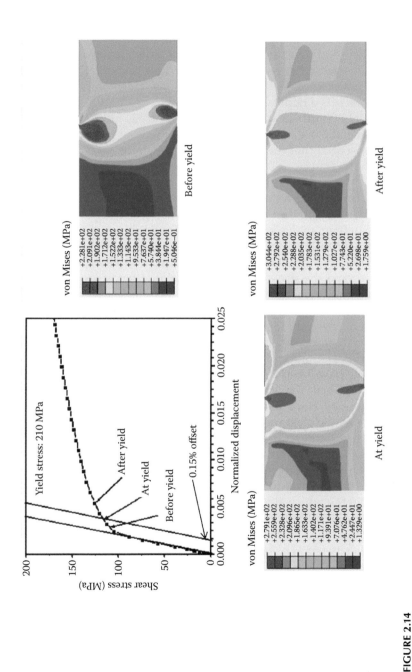

FIGURE 2.14

Von Mises stress contour profiles from FE analysis of shear punch test at instances before yield, at yield (0.15% offset), and after yield for AISI 316. Yielding is indicated by the fully developed through-thickness plastic zone.

2.3.3 Effect of Specimen Thickness and Clearance Zone

The load-displacement data of the shear punch test is an interrelated function of the punch diameter, punch-die clearance width (c), specimen thickness (t), and thickness-to-grain size ratio. The punch-die clearance is an important parameter because the specimen deformation takes place in this region. The specimen thickness and punch-die clearance width are closely linked to each other as the clearance width is generally described in terms of sheet thickness in blanking operations. The literature on metal forming recommends a nominal cutting clearance of 5%–10% of specimen thickness per side (Smith 1990). When the clearance is more than the recommended value, the material undergoes deep drawing rather than shearing. It is also essential that the corners of the punch tip and the receiving die be kept sharp (radius < 0.002 mm) for obtaining repeatable data in experiments.

Guduru et al. (2007) have systematically analyzed the effects of thickness (t) and clearance width (c) on the shear yield stress using FE analysis. The specimen thickness in their analysis ranged from 0.2 to 0.8 mm and the clearance width varied from 5 to 25 μm (c/t = 2%–10%). For a fixed die-punch clearance, increasing the specimen thickness decreased the relative contribution of the bending stresses and the stress required for yielding increases. In contrast, increase in die-punch clearance at a fixed specimen thickness enhanced the bending stress and the stress required for yielding decrease. However, these systematic differences were found to be comparable to the scatter observed in the experimental measurements of the shear yield stress.

For deformation well beyond the yield stress up to the maximum stress, very few studies have been carried out to investigate the effect of clearance width on the shear maximum strength. In one such study carried out using FE analysis by Hatanaka et al. (2003) for various c/t of 2%–20%, it was shown that as the clearance width increased beyond 10% of specimen thickness, the maximum stress slightly decreased and the normalized displacement at peak stress increased. For c/t ≤ 10%, cracks generated from the vicinity of punch and die edges propagated along a straight line connecting the tips of both cracks, while for large clearances (c/t = 20%), the crack propagated only from the punch side. Such a difference in the crack propagation behavior was attributed to the stress state in the shearing region, which is affected by the clearance. Experimental results of their study also concluded that differences between the shear maximum stress for c/t of 0.02 and 0.10 were negligible.

The effect of thickness to grain size ratio (t/d_g) on the properties obtained from shear punch test data was systematically investigated by Toloczko, Yokokura et al. (2002). It was shown that for t/d_g varying between 3.1 and 18, there was no significant effect of t/d_g on the shear yield and maximum strengths of AISI 316 steels, in contrast to dependence of tensile YS on $t/d_g < 6$.

2.3.4 Tensile–Shear Strength Correlations

The shear punch technique shows real promise as it is capable of reflecting tensile behavior of materials and reproducing the metallurgical structure-property effects. The analysis of the shear punch data has been largely toward developing correlations with the uniaxial tensile data. The tensile YS–shear yield strength (τ_{ys}) correlation is linear and theoretically expressed as $YS = m_0\tau_{ys}$ where the constant m_0 is equal to $\sqrt{3}$ according to von Mises yield criterion or equal to 2.0 applying Tresca yield criterion for a stress state of pure shear. But the experimental data of many investigators revealed that the relation was found to be of type $YS = m_0(\tau_{ys} - \tau_0)$, which essentially meant that the correlation does not pass through origin, but has an offset parameter (τ_0). The values of both slope m and offset parameter τ_0 were found to depend on the alloy class.

Similar correlations were established by many investigators for UTS-shear maximum strength, where both the slope (m_1) and the offset parameter (τ_1) of the UTS correlation were higher than that of yield correlation. Table 2.1 shows the typical values of the slope and offset parameter of the yield and maximum strength correlations derived by Hamilton, Toloczko, and Lucas (1995) and Hankin et al. (1998) using a number of heat-treated microstructures of alloys such as aluminum, copper, vanadium, and stainless steel. The UTS-shear correlation for various microstructures of 2.25Cr-1Mo, mod 9Cr-1Mo, and AISI 316 steels (Figure 2.15) derived by Karthik et al. (2009) also substantiates that the slope and offset values are material specific.

The origin and significance of the material-dependent offset parameter τ_0 has never been completely understood. While few investigators attributed it to the punch-die specimen friction, the other explanation was that the limited strength range over which the data were obtained for each alloy class could not force a best linear fit through origin and hence resulted in an offset parameter τ_0. A single correlation equation with slope (m_0) of 1.8 for the

TABLE 2.1

Correlation Constants Obtained for Various Alloy Systems

Alloy	No. of Data Points	Yield Strength		Maximum Strength	
		Slope	Offset	Slope	Offset
Various aluminum	126	–	–	1.7	–8
Vanadium alloys	2	2.8	–129	1.8	–38
Stainless steel SS 316 and HT 9	10	1.7	29	2.2	–425
Cu-Zn	72	–	–	2.9	–335
Fe-Ni-Cr alloys	>30	2.0	–176	2.0	–270

Source: Hamilton, M. L. et al. 1995. In *Miniaturized Specimens for Testing of Irradiated Materials,* eds. H. Ullmaier and P. Jung, 46–59. IEA International Symposium.

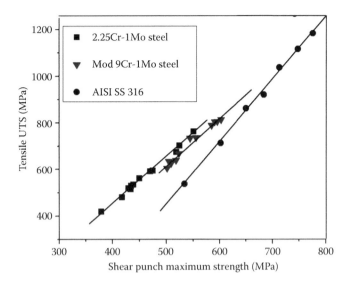

FIGURE 2.15
Linear correlation of UTS-shear maximum strength of type UTS = Aτ_{max} + B obtained for various alloys.

yield correlation and a slope (m_1) of 2.2 for UTS correlation was derived by Hamilton et al. (1995) only after subtracting the offset values (τ_0 or τ_1) from the respective data sets. Thus, the tensile-shear correlation for yield and maximum strengths has been mostly empirical in nature.

The empiricism in the correlations has been addressed by Karthik et al. (2009) (1) by modifying the experimental setup where the compliances of the experimental setup were eliminated, and (2) through improved data analysis procedures. The shear yield stress determination and its correlation with tensile YS is closely tied with the initial slope of the LDC and the use of an appropriate offset method. Eliminating the machine and fixture compliances and using the 0.15% offset definition for yield stress, Goyal et al. (2010) demonstrated that the yield correlation obtained from the experimental data of various alloys simplified into a generalized equation of type YS = 1.73τ_{ys} (Figure 2.16), which is the same as the von Mises relation between tensile yield and shear yield strengths. The experimentally obtained value of 1.73 for yield correlation constant clearly indicated that the deformation in shear punch test is shear dominant at least in the early stages of deformation.

Toward generalizing the UTS–τ_{max} relationship without involving any material specific constants, Karthik et al. (2009) made use of the analysis of Ramaekar and Kals (1986) relating the peak force in blanking to UTS and strain-hardening exponent (n) parameters. The correlation expressed as

$$UTS = m\tau_{max} \tag{2.2}$$

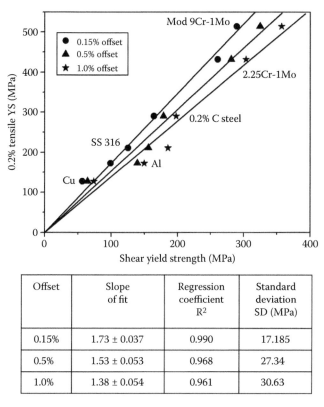

Offset	Slope of fit	Regression coefficient R^2	Standard deviation SD (MPa)
0.15%	1.73 ± 0.037	0.990	17.185
0.5%	1.53 ± 0.053	0.968	27.34
1.0%	1.38 ± 0.054	0.961	30.63

FIGURE 2.16
Linear fit between the tensile and shear yield strengths of various materials for different off-sets. The 0.15% offset definition is seen to produce the best fit of tensile-shear yield stress.

where $m^{-1} = \sqrt{\left\{\frac{1}{3}\left(\frac{3}{n}\right)^n\right\}}$ and m is a function of strain-hardening exponent (n). This equation suggests that the actual relation between the UTS and τ_{max} is a complex function related to the strain-hardening exponent of the material. This relationship was found to be obeyed (Figure 2.17) for the experimental data of various steels (2.25Cr–1Mo, mod 9Cr–1Mo, SS 316, low carbon steel) as well as nonferrous alloys (Al, Cu).

The physical significance of the UTS–τ_{max} equation could be understood as follows. The value of the coefficient m of the UTS correlation ranged from 1.0 to 1.7, depending on the n value (i.e., strain-hardening capability of the material), but always less than the yield correlation constant m_0 (= 1.73). For a brittle material whose n is low, the coefficient m is close to 1.73 (same as yield correlation constant), while for a ductile material, m reduces to around 1.10. The generalized UTS–τ_{max} correlation seems very promising, but requires prior knowledge of the strain-hardening exponent n of the material.

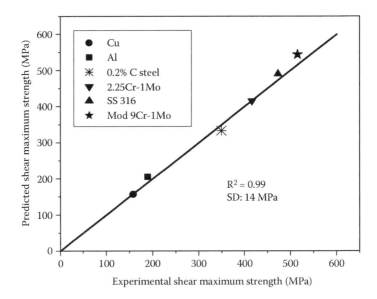

FIGURE 2.17
Plot showing the excellent agreement between the experimental shear maximum strength and that predicted using Ramaekar's equation (Equation 2.2).

2.3.5 Ductility from Shear Punch Test

In a conventional tensile test, ductility is defined by parameters such as (1) uniform elongation, defined as the strain up to the point of peak stress; (2) total elongation, defined as strain up to the point of fracture stress; and (3) reduction in area. However, in small specimen tests such as shear punch, the evaluation of the uniform or total strain directly is not possible due to the complex and inhomogeneous strain distribution across the deforming specimen and the practical difficulties in measuring the strain across a very small volume. Hence, ductility is indirectly determined from a parameter such as the strain-hardening exponent (n) by analyzing the load-displacement data. One of the earliest methods (Lucas et al. 1986) was based on two data points, τ_{ys} and τ_{max}, using the semiempirical expression

$$\left(\frac{n_\tau}{0.002}\right)^{n_\tau} = \frac{\tau_{max}}{\tau_y}. \tag{2.3}$$

This expression can be derived starting with the familiar power law Hollomon equation relating true stress (σ) to true strain (ε_p) as $\sigma = K(\varepsilon_p)^n$ (where n is the strain-hardening exponent and K is the strength coefficient) and using the well known relationship $\varepsilon_U = n$ (Dieter 1961). Linear empirical relationships between n determined from tensile stress–strain data and n_τ and uniform elongation (ε_U) – n_τ have been reported for various alloys (Toloczko, Hamilton, and Lucas 2000). A typical tensile-shear punch ductility correlation is depicted in Figure 2.18.

Alternatively, a methodology making use of the analytical models of sheet metal blanking was formulated by Karthik et al. (2011) to evaluate the strain-hardening exponent from shear punch data. The analytical models of blanking have been primarily aimed toward rapidly deriving a force-displacement graph of the blanking process, using the input values for process and material parameters and optimizing the blanking process. Recent models include those by Atkins (1980), where the blanking process is viewed purely as a shearing operation, and that of Zhou and Wierzbicki (1996), which is based on the assumption of pure tension. Klingenberg and Singh (2003) modified the Atkins model to allow for the plastic bending during the shearing process.

Combining the shear model of Atkins (1980) and the tension model of Zhou and Wierzbicki (1996) with that of Klingenberg and Singh (2005), an expression relating the strain-hardening exponent n to displacement at peak load, specimen thickness, and clearance zone width was derived by Karthik et al. (2011) as

$$t\left(\frac{n}{e}\right)^n s_f = (t - d_u)\left(\frac{1}{\sqrt{3}}\right)\left(\frac{2}{\sqrt{3}}\ln\sqrt{1+\gamma_u^2}\right)^n \Psi(d_u) \tag{2.4}$$

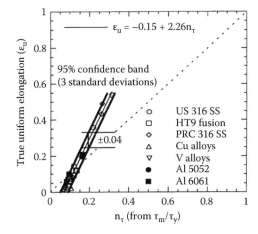

FIGURE 2.18
Correlation between tensile uniform elongation and n_τ computed using Equation 2.3 for various alloys.

where

$$\gamma_u = \frac{d_u}{c}, \; s_f = \sqrt{\frac{1}{3}\left(\frac{3}{n}\right)^n} \quad \text{and} \quad \psi(d_u) = \left(\frac{\ln\left(\frac{\pi}{2}+\gamma_u-1\right)}{\ln\sqrt{1+\gamma_u^2}}\right)^n$$

The factor $\psi(d_u)$ incorporates the corrections proposed by Klingenberg and Singh (2005) for stretching due to bending of the specimen outer fibers during shearing. Equation 2.4 gives a relationship between the parameters of the shear punch test—namely, displacement at peak load (d_u), clearance zone width (c), specimen thickness (t), and the strain-hardening exponent (n). This methodology was found to accurately predict the tensile uniform strain of steels (Figure 2.19), especially for $n \leq 0.25$, while for high strain-hardening materials like AISI 316, the model was found to underpredict the ductility. The use of this model is limited due to the assumptions of (1) plain strain conditions, (2) pure shear process with correction for stretching, and (3) the power law equation $\sigma = K\varepsilon^n$ behavior of the material. It does not include the process initiation/growth of microcracks and ductile fracture during blanking.

Since its inception, the shear punch technique has been put to a number of applications for evaluating the strength and ductility from small volumes of material by adapting simple testing geometry. Examples of applications include studies on effects of neutron and other heavy particle irradiation on

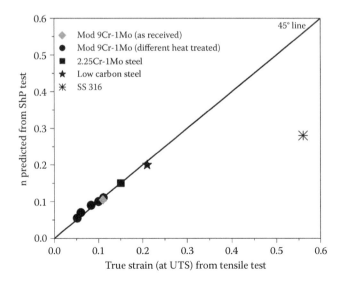

FIGURE 2.19
Plot showing the comparison of the n estimated from ShP test data using Equation 2.4 with that determined from tensile true stress–strain data.

steels and nonferrous alloys, properties in localized zones of a weld joint, deformation behavior of metallic glass, composites, etc. A few applications will be presented in Chapter 5 of this book.

2.4 Ball-Indentation Technique

The indentation-based hardness technique has emerged as a ubiquitous research tool for a number of different systems across size scales (nano to macro) and scientific/engineering disciplines. Hardness tests using spherical indenters like Brinell hardness are unique because the geometry of indentation changes with increasing penetration. In its simplest form, the ball or spherical indentation test consists of deforming a test specimen with a ball indenter under a known load. Upon removal of the indenter, a permanent impression is retained in the specimen (Figure 2.20) and the mean pressure or hardness, which is the ratio of load to the projected area of the impression, can be uniquely related to the flow stress of the material. The geometry of indentation (depth/diameter, i.e., h/d) changes with increasing penetration and hence a series of flow stress and plastic strain can be associated with spherical indentations of different sizes. This is not possible with conical or pyramidal indenters where geometrically similar indentations are produced and the ratio d/h and the flow stress remain constant with increasing penetration of the indenter.

The fundamental understanding of the spherical indentation has evolved from the works of Tabor (1951); Johnson (1970); Francis (1976); Hill, Storakers, and Zdunek (1989); and Tirupataiah and Sundararajan (1991), while various researchers, such as Haggag (1993); Oliver and Pharr (1992); Taljat, Zacharia,

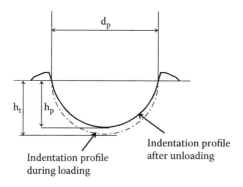

FIGURE 2.20
A typical profile obtained during a cycle of loading and unloading a metallic sample with a spherical indenter.

and Kosel (1998); Ahn and Kwon (2001); and Lee, Lee, and Pharr (2005), have established analytical and numerical procedures for accurate prediction of tensile properties from instrumented ball-indentation tests.

2.4.1 Flow Stress from Indentation

Tabor (1951) was the first to correlate the indentation hardness and the strain associated with spherical indentation to uniaxial tensile flow curve based on a premise that monotonic true stress–true strain curves obtained from tension and compression testing are similar. Tabor demonstrated that the mean pressure, H, developed between the spherical indenter and the test material increases with increasing load and can be related to the flow stress of the material as

$$\sigma = \frac{H}{\Psi} \tag{2.5}$$

where

$$H = \frac{4P}{\pi d_p^2}$$

$P = $ indentation load
$\psi = $ constraint factor
$\sigma = $ flow stress

The factor, ψ, by which the resistance to plastic flow under indentation conditions is higher than the uniaxial flow stress value is defined as the constraint factor. This constraint factor arises mainly because the plastic zone underneath the indenter is confined within a larger volume of the surrounding material, which is either elastic–plastic or rigid. The stress states in the plastic zone are hydrostatic compressive plus shear, with the shear component responsible for the plastic flow. The plastic deformation underneath the indenter is constrained, unlike in the uniaxial tensile test.

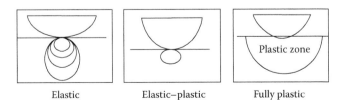

Elastic   Elastic–plastic   Fully plastic

FIGURE 2.21
Schematic of the three different stages of spherical indentation of a metallic sample with increasing loads.

For spherical indentation, the growth of the indentation geometry with increasing load occurs in three stages as shown in Figure 2.21:

1. The elastic regime, where the deformation is reversible and is described by Hertz contact solution with ψ having a constant value of 1.1

2. The transition regime, starting with the initiation of a plastic zone beneath the indenter, where the ratio H/σ increases with d_p

3. The fully plastic regime, in which the plastic zone expands to the surface of the specimen and the contact pressure increases due to work-hardening characteristics and the constraint factor attains a constant value ψ

The parameter ψ, according to Francis (1976), can be represented as a function of the parameter Φ for the three stages of spherical indentation as

$$
\begin{array}{llll}
\Psi = 1.1 & \Phi \leq 1 & \text{- Linear} \\
= C_1 + C_2 \ln \Phi & 1 < \Phi \leq C_3 & \text{- Elastic–plastic} & (2.6) \\
= \Psi^c & \Phi > C_3 & \text{-Fully plastic}
\end{array}
$$

where $\Phi = (4Eh_c)/(d\sigma)$ is a nondimensional variable and can be interpreted as the ratio of the strain imposed by the indenter to the maximum strain that can be accommodated by the material before yielding. Johnson (1970) correlated indentation data from a variety of sources and showed that Ψ is a linear function of $\ln(\Phi)$ in the transition stage of the elastic–plastic regime and transition to full plastic flow was shown to occur when the parameter Φ reaches approximately 28–30. Though slightly differing values have been presented for the parameters C_1, C_2, C_3, and Ψ^c by various researchers, such as Tabor (1951), Au et al. (1980), and Field and Swain (1995), their average values based on statistical analysis of previously published data of about 43 materials by Francis (1976) are $C_1 = 1.11$, $C_2 = 0.534$, $C_3 = 27.3$, and $\Psi^c = 2.87$.

Experimental work of Tirupataiah and Sundararajan (1991) and Matthews (1980) showed the dependency of Ψ on the strain-hardening exponent n defined by the equation $\sigma = K (\varepsilon_p)^n$, where K is the strength coefficient. Matthews derived an approximate equation for Ψ^c as a function of n as

$$
\Psi^c = \frac{H}{\sigma} = \frac{6}{(2+n)} \left(\frac{40}{9\pi} \right)^n \tag{2.7}
$$

This equation gives the value $\Psi^c = 3$ for $n = 0$ and $\Psi^c = 2.85$ for $n = 0.5$. However, the experimental results of Tirupataiah and Sundararajan for Ψ^c

investigated on iron, steel, copper alloys, and aluminum alloys lie in the range of 2.4–3.0. A larger discrepancy from Equation 2.7 was noted for materials with low strength values. It is generally accepted that the constraint factor, ψ, associated with the indentation of metallic materials by a spherical indenter is in the range of 2.8–3.0.

2.4.2 Strain Definition

The strain in spherical indentation is maximum just below the center of the region of contact and reduces with radial distance from the axis of symmetry to about 20% of the maximum value at the periphery of indentation. The average strain beneath the indenter has been defined in various ways. Tabor (1951) showed that

$$\varepsilon_p = 0.2\frac{d_p}{D} \tag{2.8}$$

Various other definitions of average strain in indentation testing have been outlined by Ahn and Kwon (2001). These include $0.4h_c/a$, $\ln[2/(1 + \cos\gamma)]$ and $0.12\tan\gamma$, where γ is the contact angle between the indenter and specimen (Figure 2.22), h_c is the actual contact depth, and a is the contact radius. Of all the strain definitions, Ahn and Kwon showed that the use of shear strain at the contact edge defined by $0.12\tan\gamma$ was found to be in best agreement between the stress–strain curves derived from ball indentation and tensile tests for various steels tested. It may be noted that the flow stress and plastic strain defined by Equations 2.5 and 2.8 respectively are representative values defined at the edge of the indentation.

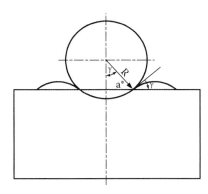

FIGURE 2.22
Schematic of typical indentation profile showing the contact angle γ.

2.4.3 Cyclic Indentation Tests

In static indentation tests, researchers had employed optical cum interferometry and profilometry techniques to directly measure the diameter of indentation from which a unique flow stress and a plastic strain can be estimated. However, such traditional methodology had limited utility for rapid measurement of the complete flow curve of metallic structures especially that in service. Based on load-depth measurement during spherical indentation of a metallic sample with progressively increasing loads at a single location, a cyclic ball-indentation test was first developed by Au et al. (1980) and then patented as the automated ball indentation (ABI) test by Haggag et al. (1990). The ABI is based on multiple indentation cycles on a metallic sample at the same location by spherical indenter. Each cycle consists of loading, partial unloading, and reloading sequences as shown in Figure 2.23. The plastic indentation depth, h_i, after complete removal of load was obtained by extrapolating the partial unloading curve linearly to zero load.

The plastic indentation diameter, d_p, is obtained using the following Hertz's relation:

$$d_p = \left\{ 0.5\,CD \left[\frac{h_i^2 + \left(\dfrac{d_p}{2}\right)^2}{h_i^2 + \left(\dfrac{d_p}{2}\right)^2 - h_i D} \right] \right\}^{\frac{1}{3}} \tag{2.9}$$

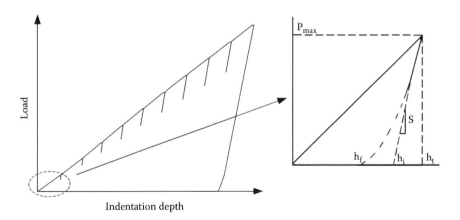

FIGURE 2.23
A typical load-indentation depth curve obtained in multiple cycle spherical indentation tests.

$$C = 5.47\,P\left(\frac{1}{E_{ind}} + \frac{1}{E_{mat}}\right)$$

where
  E_{ind} = elastic modulus of the indenter
  E_{mat} = elastic modulus of the sample

From d_p iteratively determined using Equation 2.9, the true stress and plastic strain are obtained using Equations 2.5 through 2.8, except that the Francis's constraint factor in the full plastic regime was modified by Haggag (1993) as

$$\Psi = 2.87\,\alpha_m \tag{2.10}$$

where α_m is constraint factor index and material dependent whose value varies between 0.9 and 1.25 for various structural steels depending on the strain rate and triaxial hardening.

The true stress–strain data are then fit to the familiar power law Hollomon equation:

$$\sigma = K\,(\varepsilon_p)^n \tag{2.11}$$

from which the strain-hardening exponent n is calculated as the slope of the log–log plots of true stress–strain data. This, along with the strength coefficient (K), is used to calculate the UTS using the well known expression

$$\varepsilon_U = n \tag{2.12}$$

$$UTS = K\left(\frac{n}{e}\right)^n \tag{2.13}$$

where ε_U = true plastic strain and e = 2.71.

The ball-indentation test technique has been extensively used not only for the determination of tensile properties, but also for fracture toughness. Though indentation does not produce cracking in ductile materials, methodologies have been developed for determining fracture toughness using theoretical models based on indentation energy concepts.

2.4.4 Yield Strength from Indentation

In a standard tensile test, the material deforms elastically initially; following that, plastic yielding and work hardening commence and continue in the constant volume of the specimen's gauge section until the onset of necking. In contrast, in a cyclic indentation test of loading–unloading and reloading, the elastic and plastic deformations are not distinctly separated, but take place simultaneously during the entire course of the test.

The commencement of plastic deformation (i.e., the boundary between the elastic regime and the elastic–plastic regime for each specimen) cannot be normally detected in the indentation load-depth curve as demonstrated by Ahn and Kwon (2001). The critical indentation load for the initial plastic deformation P_Y during ball indentation is obtained as

$$P_Y = \frac{R^2 \pi^3}{6E_r^2} (1.5\ P_m)^3 \tag{2.14}$$

where E_r is the reduced modulus and P_m is equal to 1.1 times the yield strength YS at the initial plastic deformation. For example, the value of P_Y for AISI 1025 steel is calculated as 0.2 gf by setting $E = 160$ GPa, $n = 0.3$, $E_{ind} = 400$ GPa, $v_{ind} = 0.28$, $R = 0.5$ mm, and YS $= 196$ MPa, which is too low to be detected in usual ball indentation systems. Therefore, an accurate determination of yield strength in an indentation test should be based on the entire load-depth curve of the test.

A different approach (George, Dinda, and Kasper 1976) is used for the estimation of yield strength from ball indentation test using the Meyers relationship:

$$\frac{P}{d_t^2} = A \left(\frac{d_t}{D} \right)^{m-2} \tag{2.15}$$

where d_t is the total diameter of the residual impression, m is the Meyers exponent, and A is the material constant. The yield strength (YS) is related to the parameter A as

$$YS = \beta_m\ A \tag{2.16}$$

where β_m is a constant for a given class of materials. The value of β_m for each class of materials is determined from tensile yield strength and the value A obtained from a ball indentation test. A value of $\beta_m = 0.22$ for carbon steels and 0.191 for stainless steels has been reported (Haggag et al. 1990).

2.4.5 Contact Area and Pileup/Sink-In Phenomena

Since the stress–strain values are defined on the basis of contact area between the indenter and material in the loaded state, accurate determination of contact area is essential for an accurate evaluation of stress and strain. For a typical load-depth curve shown in Figure 2.23, Oliver and Pharr (1992) showed that the elastic depression of the circle of contact (Figure 2.20) below its unloaded position is given by

$$h_e = q \, P_{max} \, dh_e/dP \tag{2.17}$$

where q is 0.75 for a paraboloid of revolution and dh_e/dP is the inverse slope of the initial unloading line (Sneddon 1965). The contact depth considering only the elastic deflection is expressed as

$$h_c = h_{max} - h_e \tag{2.18}$$

Defining the inverse unloading slope as $(h_{max} - h_i)/P_{max}$, the contact depth simplifies as

$$h_c = h_{max} - 0.75 \, (h_{max} - h_i) \tag{2.19}$$

where h_i is the intercept depth obtained by extrapolating the tangent line of the initial unloading curve to P = 0. This equation describes the contact depth after considering the elastic deflection depth only. Using the geometry of ideal spherical configuration, the contact diameter d_p can be determined from h_c as

$$d_p = \sqrt{h_c D - D^2} \tag{2.20}$$

where D is the indenter diameter.

The plastic indentation diameter evaluated from Equation 2.9 or 2.20 are with reference to the original plane of the specimen surface. They do not factor an important phenomenon that comes into play during indentation namely the pileup or sink-in behavior. The raising up/dipping in of the circle of contact relative to the original surface is referred to as the pileup/sink-in phenomena. The piling-up or sinking-in effects during indentation significantly influence the actual indentation diameter and thereby the actual contact area.

Norbury and Samuel (1928) measured the contact circle diameter on the recovered indentation and calculated its height relative to the original surface. They found that the ratio of the height of pileup to penetration depth (i.e., h_{pile}/h_c; see Figure 2.24) is dependent on n, the strain-hardening exponent of the indented material. Hill et al. (1989) provided an analysis for pileup/sink-in based on deformation theory using a parameter, c^2, expressed as

$$c^2 = \frac{a^2}{h_c^* D} = \frac{h_c}{h_c^*} = 1 + \frac{h_{pile}}{h_c^*} = \frac{a^2}{a^{*2}} \tag{2.21}$$

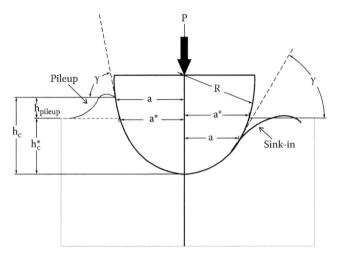

FIGURE 2.24
Schematic of the deformation around the indenter showing the pileup and sink-in during spherical indentation.

where
  c is a numerical invariant
  a* is the contact radius in the plane of original surface
  a is the contact radius in the plane of pileup/sink-in

In this equation, the last approximate relation is derived from geometric relation in a shallow indentation ($a \ll D/2$). The parameter c^2 is interpreted as the ratio of the depth along which contact is made to the total depth of penetration. Thus, $c^2 < 1$ implies that material sinks in, while $c^2 > 1$ means that material piles up. Fitting this invariant c to the experimental data obtained by Norbury and Samuel (1928) for a range of materials, Matthews (1980) suggested an empirical relation:

$$c^2 = \left(\frac{1}{2}\right)\left(\frac{2+n}{2}\right)^{\frac{2(1-n)}{n}} \tag{2.22}$$

Hill et al. (1989) proposed an analytical relationship between c^2 and n using finite element analysis (FEA) based on deformation theory as

$$c^2 = \frac{a^2}{a^{*2}} = \frac{5}{2}\left(\frac{2-n}{4+n}\right) \tag{2.23}$$

$c^2 = 1$ marks the boundary between the piling up and sinking in, which, according to Equation 2.23 corresponds to a strain-hardening exponent (n) of about 0.28.

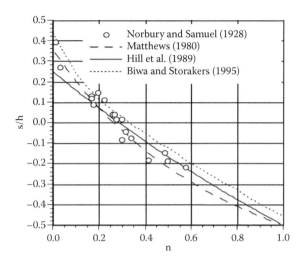

FIGURE 2.25

Experimental data of Norbury and Samuel (1928) superimposed on data from models developed by Matthews (1980), Hill et al. (1989), and Biwa and Storakers (1995), showing the dependence of the pileup parameter on the strain-hardening exponent, n.

Biwa and Storakers (1995) noted that the actual subindenter deformation state is not proportional and analyzed the problem using FEA based on incremental theory of plasticity. The results were quantitatively similar with slightly different numerical predictions for pile up and sink in. The predictions of the theoretical models of Hill et al. (1989) and Biwa and Storakers (1995) overlapped on the experimental data obtained by Norbury and Samuel (1928) for a variety of metals and the empirical relation of Mathews as shown in Figure 2.25.

2.4.6 Numerical Studies on Ball Indentation

With advances in computational mechanics in the last two decades, rapid strides have been made in the numerical analysis of the nonlinear elastic–plastic problems such as spherical indentation using finite element methods. There has been a lot of research work on methods to predict the elastic–plastic material properties The FE-based methods have been extensively employed to examine the various aspects of spherical indentation such as (1) effects of indenter-specimen friction, (2) compliance of indenters/machines, (3) stress–strain distribution in the deformed volume, and (4) pileup/sink-in behavior, etc.

2.4.6.1 *Machine Compliance*

The evaluation of compliance of the indentation test machine is very essential for accurate determination of sample deformation. In load frames of finite stiffness, the load applied via the indenter induces displacement in both the sample and the load frame, and the machine compliance modifies the indentation response. For example, in a typical ball-indentation test setup shown in Figure 2.26, the indentation depth measured includes the elastic deflection of the indenter holder up to the point where the displacement sensor, such as LVDT, is mounted. Thus, the elastic deflections of the experimental setup must be identified and subtracted from the total indenter displacement to account properly for sample deformation. The elastic deformation of the indenter is also a component of this compliance and is negligibly very small if the elastic modulus of the indenter is at least three times higher than that of the indented material, such as a tunsgten carbide indenter for steel samples or steel balls for aluminum samples.

The methodology proposed by Doerner and Nix (1986) for computing the frame compliance is briefly explained here. The total compliance (C_{total}) is the sum of the sample compliance (C_s) and frame compliance (C_f) in series:

$$C_{total} = C_s + C_f \qquad (2.24)$$

The total compliance (C_{total}) is dh/dP, determined as the inverse slope of the tangent line of the experimental unloading curve at the maximum loading point. The specimen compliance is computed according to Sneddon's (1965) elastic punch solution relating the reduced elastic modulus of the sample E_r to the projected contact area (A) as

$$C_s = \frac{1}{2E_r}\sqrt{\frac{\pi}{A}} \qquad (2.25)$$

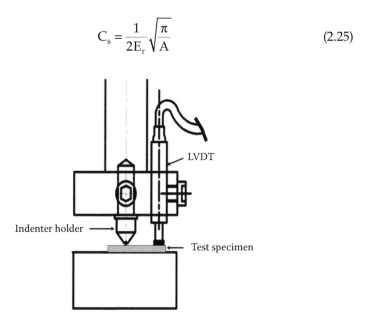

FIGURE 2.26
Schematic showing the positions of the indenter holder and the LVDT in a typical ball indentation experimental setup.

$$E_r = \left(\frac{1 - v_{ind}^2}{E_{ind}} + \frac{1 - v_{mat}^2}{E_{mat}} \right)^{-1} \tag{2.26}$$

where E_{ind} and v_{ind} are the elastic modulus and Poisson's ratio of the indenter, while E_{mat} and v_{mat} are the elastic properties of the indented material.

Substituting Equation 2.25 for C_s, Equation 2.24 can be rewritten as $y = mx + b$, where y is the total compliance (C_{total}) and x is $1/\sqrt{A}$. Computing C_{total} from a set of experimental data (P_{max}, h_{max}), a linear regression of $C_{total} - 1/\sqrt{A}$ plot yields machine compliance C_f. The load-depth data of the material are then corrected for the machine compliance as

$$h_{t(corr)} = h_t - C_f P \tag{2.27}$$

The C_f reported by various investigators are typically around 10^{-6} mm/N.

2.4.6.2 Friction Effects

Friction at the indenter–specimen interface plays a crucial role in the spherical indentation process and the difficulties associated with experimental evaluation of friction and its potential variability during indentation underpins the need for understanding their effects. The friction coefficient (f) for two metals in contact is usually in the range of 0.1–0.4. For a diamond indenter with a steel surface, it has been shown that f ~ 0.1–0.15, while for tungsten carbide to steel contact, f ~ 0.4–0.6. The effects of friction have generally been investigated through finite element analyses using the Coulomb friction model.

Friction is known to influence the contact conditions and distribution of plastic strain in the subindenter region. The analysis of Mesarovic and Fleck (1999) revealed that friction simultaneously promoted yielding processes in the region beneath the indenter and also constrained the lateral spreading of plastic flow. Most of the studies (Cao, Qian, and Huber 2007; Beghini, Fontanari, and Monelli 2009; Karthik et al. 2012) suggest that the effects of friction on the contact conditions are strongly felt when the friction coefficient (f) takes low values, while a saturation is reached when f reaches medium to high values. The slope of the FE-simulated load-depth curve (Figure 2.27), normalized as P/D^2 and h_t/D, respectively, is seen to be affected by friction for greater penetration depths, with the slope of the curve increasing with increasing friction coefficient; however, a tendency toward saturation is observed for f > 0.2.

Friction affects the indentation profile in a manner that depends on strain-hardening exponent, n. A frictionless contact (f = 0) for material having high plasticity and a low strain-hardening exponent (n) leads to material flow

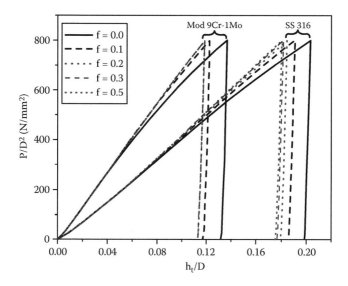

FIGURE 2.27
Influence of friction coefficient (f) between indenter and specimen on the load-depth plot of BI test on two different steels.

more around the indenter, leading to significant pileup. The frictional effects on the extent of pileup, as depicted in Figure 2.28, are dominant for f = 0 and 0.1, but the profiles tend to converge at f = 0.2 and above. On the other hand, for strongly hardening material (n > 0.3), there is no discernible effect of friction on the indentation profiles at the contact edge. Frictional effects on pileup are most significant in materials that do not strain harden (low n) and nonzero frictional value acts to reduce the pileup.

2.4.6.3 Analysis of Pileup/Sink-In

Equation 2.23 relating the extent of pileup to strain-hardening exponent (n) provides a means by which the indentation diameter and contact area can be corrected for the influences of pileup. This, however, requires prior knowledge of n. An iterative process suggested by Ahn and Kwon (2001) starts with an arbitrary input of n for computing the corrected contact area and hence a flow stress–strain. The value of n obtained from the flow stress–strain relation is compared with input n and the iterative process is continued till the input n equals the output value. Using this iterative procedure, it is seen that true stress–strain evaluated from ball indentation (BI) with pileup corrections compared well with a tensile stress–strain curve as seen in Figure 2.29. This plot highlights the importance of pileup corrections in stress–strain evaluation for materials with *n* < 0.2, wherein neglecting pileup corrections can result in errors as high as 20% in calculated material properties.

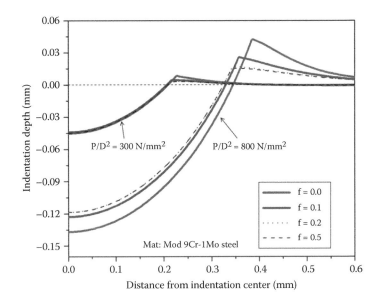

FIGURE 2.28
Indentation profiles in the loaded state of mod 9Cr-1Mo steel from FE for various friction values and two different indentation loads.

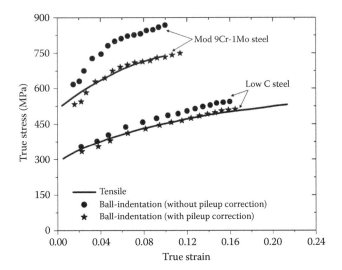

FIGURE 2.29
Stress–strain data from BI test superimposed on the tensile stress–strain data for mod 9Cr-1Mo and low carbon steels.

Based on finite element simulations of indentations on many materials with different n, Taljat et al. (1998) proposed two expressions for c^2 for fully loaded and unloaded indents:

$$\text{Loaded } c^2 = \frac{1}{4}(5 - 3n^{0.7}) \tag{2.28}$$

$$\text{Unloaded } c^2 = \frac{1}{10}(13 - 8.5n^{0.8}) \tag{2.29}$$

Equations 2.22, 2.23, 2.28, and 2.29 for pileup corrections are based on fully plastic analyses and assume that the c^2 value is solely dependent on n. Habbab, Mellor, and Syngellakis (2006) showed that, in addition to n, c^2 is also dependent on indentation depth h for low and moderate indentation depths (0.01D < h < 0.06D) and they presented modified equations for pileup corrections. However, c^2 gets closer to that predicted by Equation 2.23 for deeper indentation in the fully plastic regime. During the elastic–plastic transition, pileup (c^2) grows in a way that depends on the relative amount of elastic and plastic deformation and was characterized by Taljat and Pharr (2004) by a nondimensional unified parameter $(E/\sigma_y)(h_c/a_c)$. It was shown that the pileup becomes constant only when the parameter $(E/\sigma_y)(2h_c/a_c) > 1000$, well beyond the value $(E/\sigma_y)(2h_c/a_c) > 100$ that marks the beginning of the fully plastic regime. These studies indicate that pileup and sink-in behavior is a complex phenomenon that is not amenable to simple analytical descriptions.

2.4.6.4 Numerical Methods for Stress–Strain Evaluation

There have been a lot of attempts to predict the material elastic–plastic properties based on reverse engineering using numerical methods. These methods are based on analysis of load-indentation depth curve (P–h curve) or crater profile and numerically mapping the indentation variables (P, h, d) to the elastic and plastic properties (E, YS, n, σ, ε_p). In the method developed by Beghini et al. (2006), a database of load (P) versus depth of penetration (h) curves is generated for various combinations of E, YS, and n using finite element simulation. The load-depth data of a single loading–unloading test is fitted to a combination of elastic–plastic parameters to derive a functional relationship involving about 200 fitting coefficients (A_k),

$$\frac{P^{FE}}{E_s D^2} = \sum_{k=1}^{4} A_k \left(\frac{h^{FE}}{D} \right)^{k/2} \tag{2.30}$$

The best fitting coefficients A_k are calculated for any combination of the parameter σ_y and n used by means of a two-dimensional polynomial function using the least square fitting method:

$$A_k = \sum_{i=1}^{6} \sum_{j=1}^{6} \alpha_{ijk} \sigma_y^{i-1} n^{j-1}, \ k = 1, 2, 3, 4 \tag{2.31}$$

The set of constant coefficients, ijk, for one class of material constitutes the database.

By adopting an optimization algorithm, the material parameter, σ_y, and n are deduced by minimizing the function

$$\chi(E, \sigma_y, n) = \sum_{m=1}^{M} \left[P_m^{FE} \left(h_m^{Exp}, \sigma_y, n \right) - P_m^{Exp} \right]^2 \tag{2.32}$$

where M number of data points are chosen on the $P^{Exp} - h^{Exp}$ curve. The optimization algorithm scans the domain of material properties (σ_y, n), and selects the theoretical curve $P^{FE} - h^{FE}$, which minimizes the function $\chi(E, \sigma_y, n)$, and the corresponding σ_y and n are the predicted material properties. Using this procedure, Beghini, Bertini, and Fontanari (2006) could predict the material properties with a relative error of within 4% for σ_y and within 2% for n. It may be noted that the indentations were limited to shallow depths up to a maximum nominal pressures of P/D^2 of 250 MPa or h/D of 0.04.

While most of the indentation theories were based on the deformation theory of plasticity, which was valid only for proportional loading cases, Lee et al. (2005) analyzed the indentation using incremental plasticity-based FE analyses. Their analysis, based on incremental plasticity, shows that the maximum equivalent stress occurs at the surface part 0.4d away from the indentation center, while it occurs at the bottom part of the indentation center for deformation theory. Based on the stress–strain distribution of the subindenter, Lee et al. (2005) proposed a new optimal point (instead of indentation edge) at 10% beneath the indenter diameter from the surface at a distance 0.4d in the radial direction, with negligible frictional effect and strain gradient for stress–strain formulation. Functional relationships for c^2, ε_p, and ψ were derived in terms of normalized indentation parameters (h/D, n) and a numerical procedure was developed to compute E, YS, and n from a given P–h curve. The material properties were evaluated at this reference point using regression formulae of indentation variables c^2, e_p, and Ψ. As in Beghini and colleagues' method, the indentation depth was limited to 6% of the indentation diameter (D).

Mapping functions in indentation are truly complicated; alternatively, the concept of representative strain (Cao and Lu 2004), which is defined based on dimensional analysis, and large deformation FEM have been attempted to analyze the indentation response. Using the Π theorem of dimensional analysis, the indentation load P, expressed as a function of (E, σ_r, n, h, R), can be rewritten as

$$P = \sigma_r h^2 \Pi\left(\frac{E^*}{\sigma_r}, n, \frac{h}{R}\right) \tag{2.33}$$

where σ_r is the representative stress and E^* is the reduced modulus.

For a given indentation depth h_g and an indenter radius R, the preceding equation reduces to

$$\frac{P}{\sigma_r h_g^2} = \Pi\left(\frac{E^*}{\sigma_r}, n\right) \tag{2.34}$$

where the dimensionless function Π relates the indentation response to the material properties. The problem is that of identifying a strain level that makes the dimensionless function independent of the strain-hardening exponent and only dependent on the parameter E^*/σ_r. Based on FE computation, Cao and Lu (2004) derived an expression for dimensionless function Π independent of n for a representative strain of $\varepsilon_r = 0.03$ as

$$\Pi\left(\frac{E^*}{\sigma_r}\right) = C_1 \ln^3\left(\frac{E^*}{\sigma_r}\right) + C_2 \ln^2\left(\frac{E^*}{\sigma_r}\right) + C_3 \ln\left(\frac{E^*}{\sigma_r}\right) + C_4 \tag{2.35}$$

where the coefficients C_1, C_2, C_3, and C_4 are dependent on the parameter h_g/R. The representative strains ε_r corresponding to various indentation depths are numerically identified as a function of (h_g/R) using FE computation as

$$\varepsilon_r = 0.00939 + 0.435\left(\frac{h_g}{R}\right) - 1.106\left(\frac{h_g}{R}\right)^2 \left(0.01 < \frac{h_g}{R} < 0.1\right) \tag{2.36}$$

By performing indentation tests to two different depths (e.g., $h_{g1} = 0.01R$ and $h_{g1} = 0.06R$), representative stresses σ_{r1} and σ_{r2} are computed using Equations 2.34 and 2.35 and derived coefficients C_1–C_4. Cao and Lu (2004) demonstrated that σ_y and n could be uniquely determined using the indentation loads at two different indentation depths using the expression

$$\sigma_r = \sigma_y\left(1 + \frac{E}{\sigma_y}\varepsilon_r\left(\frac{h_g}{R}\right)\right)^n \tag{2.37}$$

They demonstrated validity of this procedure for materials with $700 \geq E/\sigma_y \geq 65$ and $0 < n < 0.5$.

Citing the limitations of this method—namely, lower penetration depths and applicability over limited range of E/σ_y, Zhao et al. (2006) developed a new algorithm on a similar representative strain concept, but with higher penetration depths of h_g/R of 0.13 and 0.3 and additional dimensionless function Π_1 involving the contact stiffness $S = dP/dh$ (obtained from the slope of the initial portion of the elastic unloading curve) as

$$\frac{S}{E^*} = \Pi_1 \left(\frac{E^*}{\sigma_r}, n, \frac{h}{R} \right) \qquad (2.38)$$

The material parameters in their study varied over a large range to cover essentially all engineering materials, with E/σ_y from 2 to 3000 and n from 0 to 0.6. The dimensionless functions proposed in their work were dependent on both the ratio of the reduced Young's modulus (E^*) to the representative stress and the strain-hardening exponent; however, friction effects were not included.

Most of these inverse methodologies based on analysis of the P–h curve were limited to shallow indentation tests with $h_{max} \leq 0.1R$, probably to minimize the contribution of specimen–indenter friction to the load-depth curve and thereby exclude the friction parameter in the analysis procedure. But low h_{max}/R indentation is weak in the uniqueness issue (i.e., for shallow indentation, two materials with different properties could produce quite similar load-depth curves, making the data inversion scheme less effective). On the other hand, with large h_{max}/R, friction effects come into play and must be included in the analysis. This will require higher computational time due to increase of FEA simulations and postprocessing job.

Cao et al. (2007) in later work investigated the effect of friction on the parameter h/R and proposed several novel methods to extract the mechanical properties of power law engineering materials based on indentation-response-based definitions of the representative strain and energy-based concepts. Lee, Kim, and Lee (2010) extended previous work of mapping the P–h curve to the σ–ε curve for deeper indentation tests up to $h_{max} \sim 0.40R$, using modified regression functions between indentation parameters and load-depth data numerically evaluated. An optimal data acquisition point at $2r/d = 0.8$ and $h/D = 0.3$, where the strain gradient is gentle and frictional effect is negligibly small, was identified for the numerical procedure. The average errors of evaluated material properties were less than 5% for power law hardening materials.

Artificial neural network (ANN) models have also been attempted to predict material properties from indentation load-depth curve. ANN has the capability to identify the underlying functional relationship in the data, and their ability to learn by example makes them very flexible and powerful. The basic units of neural networks are the artificial neuron inspired by the biological nervous system. Neural networks analyze data by passing it through several simulated processors that are interconnected and highly distributed. Neural network processes by accepting inputs, x(n), which are then multiplied by a set of weights, w(n). The neurons then nonlinearly transform the sum of the weighted inputs, by means of a transfer function, f, into an output value α, as shown in Figure 2.30. The transfer function is determined largely by the connections between elements. The prediction of material properties by indentation tests using neural network models is made possible by training, tuning, and validating the network with a pair of input and output data obtained from either experimental or computational data.

The scheme for material property prediction using ANN is shown in Figure 2.31. A database of load depth is generated using finite element analysis by numerically simulating the indentation testing for various combinations of yield strength, strain-hardening exponent (n), and strength coefficient (K). The input is provided to train the ANN. The neural network employed by Sharma et al. (2011) for data inversion makes use of a multilayer perceptron (MLP) network with a back propagation algorithm. It consists of six layers before the output stage and five layers consisting of the intermediate neurons also known as hidden layers. Their model was capable of predicting the n and K values for steels within 3%–4% of the tensile test results.

It is evident that the identification of plastic properties with spherical indentation has been a subject of rigorous studies in the last decades. Though spherical indentation has experimental simplicity, the stress and strain fields under an indenter are complex and, even for bulk isotropic materials, analysis of data is nontrivial. The discussions here have been restricted to bulk metallic systems from a purely continuum perspective. For more complicated materials systems (e.g., thin films, small volumes, porous structures,

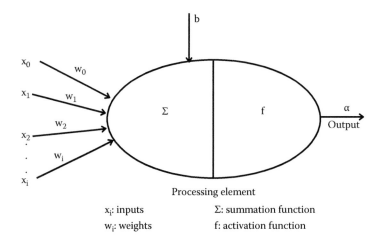

FIGURE 2.30
Schematic of an artificial neural network.

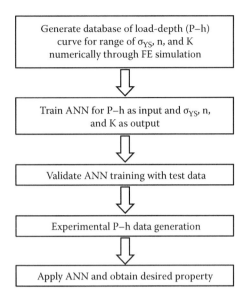

FIGURE 2.31
Methodology for evaluation of stress–strain from ball indentation load-depth data using ANN. (From Sharma, K. et al., *Sadhana* 36:181–192, 2011.)

biomaterials), indentation response is tied to specific aspects of material behavior, yet effective interpretation requires expertise in both deformation mechanics and the underlying physics of the system being indented.

2.5 Small Punch Test

The small punch (SP) technique has emerged as one of the versatile small specimen test methods for evaluating tensile, fracture, and creep properties using specimens typically of about 3–8 mm diameter and 0.25–0.5 mm thickness. Also referred to as the ball punch technique, the test method involves centrally loading a clamped disk sample using a ball into a larger size cavity of the lower die up to failure, causing a bulge in the deformed specimen.

This technique was a spin-off from the miniature disk bend test (MDBT) originally developed by Manahan et al. (1981) and Huang, Hamilton, and Wire (1982) for quick assessment of ductility loss in neutron-irradiated steels to aid in screening candidate alloys for nuclear applications. The specimens in disk bend tests are simply supported and centrally loaded by a spherical punch. The load-punch displacement curve was interpreted by approximating the deformation as symmetric bending of a circular plate and the strain (ε) generated at the center of a circular disk was derived as

$$\varepsilon = \frac{tw}{(a^2 + t^2)} \tag{2.39}$$

where
a is the disk radius
t is the specimen thickness
w is the central deflection

This equation based on elastic behavior is applicable only for small strains (<5%). Later, the technique evolved as a small punch test (Okada, Lucas, and Kiritani 1988); the only difference was that the specimen was clamped along its perimeter. In addition to flow properties, the LDC of small punch loading configuration produced a definite ductile–brittle transition with decreasing test temperature similar to the conventional FATT (fracture appearance transition temperature) of conventional Charpy impact tests. In the following sections, the focus will be on tensile flow properties from SP tests, while the fracture-related properties from SP techniques will be dealt with in the next chapter.

The SP test employs a spherically tipped punch and is different from shear punch configuration, which employs a flat-ended cylindrical punch for shearing the sample into a receiving die. The specimen deformation is not restricted to a localized process zone as in shear punch, but is spread over the whole region of the specimen that is not clamped. The characteristic dimensions of the small punch test setup (Figure 2.32) are ball diameter (d), hole size (D), specimen thickness (t), and die radius (R). The receiving die hole size is generally designed per the relation D ≥ d + 2t to ensure that the

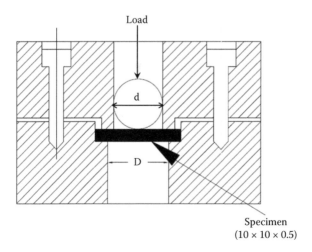

FIGURE 2.32
Loading configuration of small punch test.

deformation is close to bulging and is not subjected to frictional forces arising from contact of the specimen with the inner wall of the receiving die.

2.5.1 Deformation Regimes in Small Punch Tests

The stress state is approximately biaxial tension; however, the stress and the
deformation states change with radial position of the bulged specimen. A
typical LDC recorded during the test exhibits distinct deformation regimes,
as shown in Figure 2.33. The first regime corresponds to the elastic bending, during which the entire sample undergoes elastic deformation. During
the second regime—namely, the plastic bending—the plasticized volume in
the center of the sample progressively increases, spreading from the contact
zone of the sample with the ball to the overall thickness and also spreading
in the radial direction. The subsequent inflection point is associated with
a transition from bending to membrane stretching, finally leading to load
maximum associated with local thinning and crack initiation (plastic instability). The increasing contact of the ball with the specimen with increasing
penetration in small punch results in a plastic membrane stretching regime.
It may be noted that this regime does not exist in a shear punch test as it is a
constant area loading configuration.

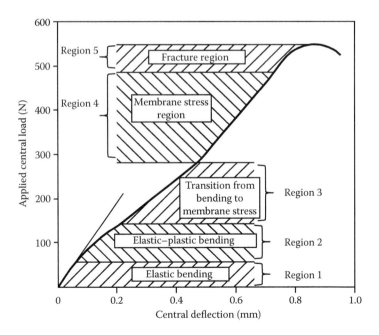

FIGURE 2.33
Typical load-displacement plot obtained in small punch test with the different deformation
regimes indicated.

The nature of the LDC is evidently dependent on the ball diameter (d), receiving die hole diameter (D), and specimen thickness (t). Various specimen sizes have been employed by different researchers for SP tests, ranging from TEM disk specimens of 3.0 mm diameter and 0.25 mm thickness to large specimen sizes of 8.0 mm diameter/10 × 10 mm² and 0.5 mm thickness. The corresponding punch and receiving die hole diameter combinations (d, D) range from (1.0, 2.5 mm) to (2.5, 4.0 mm).

With increase in specimen thickness, the applied load is higher, but the overall shape of the LDC remains the same. For a given specimen thickness, the inflection point corresponding to transition from bending to membrane stretching occurred at small displacements and higher loads as d increased and D decreased. The maximum load increased with increased d and t. These trends (Lucas 1990) reflect the general effects of the specimen thickness and punch/die dimensions on the resulting curve (Figure 2.34). For a fixed D and d, the effects of specimen thickness (varying from 0.4 to 0.5 mm) on yield load P_y and peak load P_m of the SP test on a steel sample are shown in Figure 2.35 (Matocha 2015).

The use of simple plate bending theory to study the initial deformation in small punch has limitations because of local indentation beneath the ball. The plastic zone is formed locally beneath the ball due to shear stress generated by indentation in addition to the bending stress. By analyzing the stress state of the disk bend deformation by simple analytical models, Byun et al. (2001) deduced the stress components for both top and bottom surfaces of the disk bend specimen as a function of the deflection (Figure 2.36). The

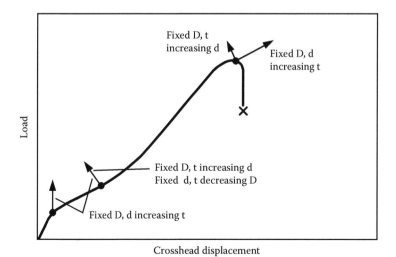

FIGURE 2.34
Plot showing the effect of specimen thickness and die dimensions on SP load-displacement curve.

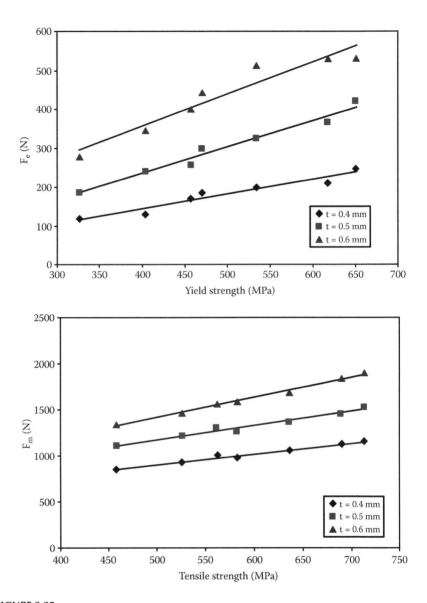

FIGURE 2.35
Plots showing the dependence of specimen thickness on SP yield load–YS and SP peak load–UTS correlations.

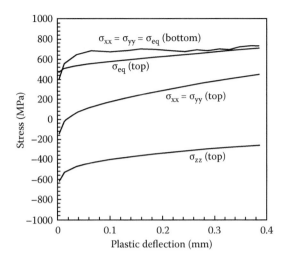

FIGURE 2.36
Plot showing the principal stress components at top and bottom surfaces of the disk bend specimen as a function of the deflection in an SP test on SS 316 LN steel.

principal stress components at the center of the top surface in the loading (z) direction are compressively defined by the equation

$$\sigma_{zz} = -\frac{4P}{\pi d_p^2} \tag{2.40}$$

The perpendicular components and equivalent stress are tensile and defined by the equations

$$\sigma_{xx} = \sigma_{yy} = \sigma_{zz} + \sigma_{eq} \tag{2.41}$$

$$\sigma_{eq} = \frac{1}{\sqrt{2}} \left((\sigma_{xx} - \sigma_{yy})^2 + (\sigma_{yy} - \sigma_{zz})^2 + (\sigma_{zz} - \sigma_{xx})^2 \right)^{\frac{1}{2}} \tag{2.42}$$

The principal stress components in the perpendicular directions (x and y) change from compressive to tensile when the disc deforms by stretching. At the center of the bottom surface, $\sigma_{zz} = 0$ and the stress state is biaxial with equal x and y stress components. The triaxiality defined as the ratio of mean stress to equivalent stress at the top surface is negative initially and increases with deflection, while that at the bottom of the specimen surface is constant (~2/3) close to the value in a necked region of a tensile specimen.

2.5.2 Correlation with Tensile Properties

Due to the complexity of the stress state developed during the SP test, initial development focused on establishing an empirical relationship between the different mechanical properties and certain characteristic points of the SP curve, such as the yield load P_y, peak load P_m, and the corresponding peak displacement d_m.

The deviation of the initial portion of the curve from linearity is normally designated as yield load. The SP yield load (P_y) of various specimen thickness (t) can be scaled as P_y/t^2, where t is the initial thickness of the disk, while peak load (P_m) could be normalized for various specimen thicknesses as $P_m/(t \times d_m)$, where d_m is the displacement corresponding to peak load (Mao and Takahashi 1987; Hurst and Matocha 2010). Unlike the shear punch test, where the curves of different specimen thicknesses are scaled as P/t, the normalizing factor in the small punch test is approximately P/t^2.

The yield stress and maximum stress correlation are expressed as

$$YS = \alpha \frac{P_y}{t^2} \qquad (2.43)$$

$$UTS = \beta \frac{P_m}{(t \cdot d_m)} \qquad (2.44)$$

where α and β are parameters related to the tested material. A typical correlation (Matocha 2015) for low alloy steel is shown in Figure 2.37 and the correlations are shown to be independent of specimen thickness according to Equations 2.43 and 2.44.

2.5.3 Numerical Studies

It is clear that the flow behavior in small punch tests is much more complicated than those of conventional tests, and evaluating the material true stress–strain relationship from SP tests is not straightforward. The main challenge is the nonhomogenous stress and strain fields in the deforming specimen making it difficult to evaluate the material parameters. A number of researchers have employed FE analysis to interpret the deformation and derive solutions for flow properties.

The stress and strain fields analyzed (Cheon and Kim 1996) using FE simulations revealed significant differences in the initial deformation behavior of 0.25 and 0.5 mm thick specimens for the same punch diameter and loading configuration. The increase in specimen thickness makes the initial LDC deviate from linearity at higher loads. In the case of a 0.5 mm specimen, the shear stress beneath the indenter was dominant and the deviation from linearity was due to loss of constraint by the surrounding material, while for the 0.25 mm specimen, radial propagation of plastic bending influences the initial non linearity. The maximum plastic strain for the 0.5 mm specimen

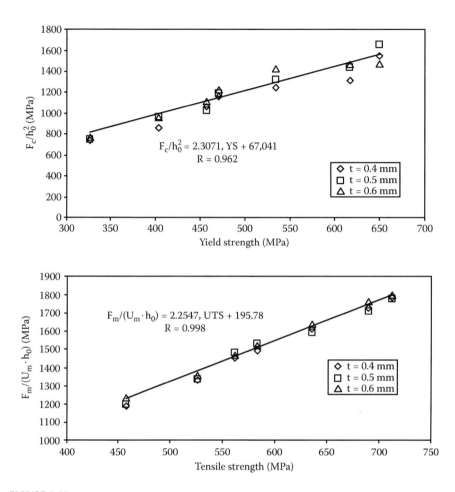

FIGURE 2.37
Plots showing the dependence of specimen thickness on SP yield load–YS and SP peak load–UTS correlations.

was at the edge of contact, while for the 0.25 mm specimen, plastic strain was peaking at the center of specimen–ball contact. Similar studies by Campitelli et al. (2004), with specimen thickness of 0.15, 0.25, and 0.35 mm, showed that the shapes of the LDC were not self-similar and the entire LDC cannot be normalized by $(t_0)^2$. The separation of plastic bending and membrane stretching regimes was more pronounced when a well dominated elastic bending regime occurred in thinner samples. The thinner samples were seen to show microyielding in tension near the bottom surface of the disk, while the thicker ones showed yielding in compression near the top surface with the dominant indentation regime.

FE analysis has aided in studying the sensitivity of the LDC to factors such as initial clamping force and friction at the ball–specimen interface. A

typical FE model of small punch configuration shown in Figure 2.38 consists
of specimen, punch, die, and specimen holder, where the latter three components are modeled as rigid bodies. The test specimen is generally modeled
as isotropic elastoplastic material obeying the von Mises yield criterion and
associative J_2 flow theory. Simulating the initial clamping force by displacing
the upper die toward the lower die to different levels, the resulting load-
displacement curves (Figure 2.39) indicate negligible effect of initial clamping force. The friction coefficient at the ball specimen was found to have
minimal effect on the initial path followed by LDC but significantly affected
the peak load as seen from Figure 2.40, generated for AISI 304 stainless steel

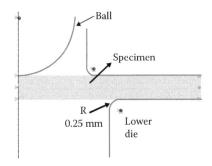

FIGURE 2.38
Schematic of the axisymmetric finite element model of an SP test.

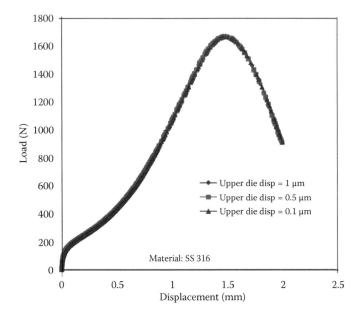

FIGURE 2.39
Effect of initial clamping force on the LDC of an SP test.

for three friction coefficients f = 0, 0.1, and 0.2. With an increasing friction coefficient, the point of peak plastic strain is seen to shift away from the center of the specimen (Figure 2.40).

The strains achieved in small punch are very large compared to that achieved in a tensile test. This has implications in validation of FE simulation with tensile data as input. The results of comparison of FE-generated LDC with experimental curves for two steels—namely, AISI 304 and modified 9Cr–1Mo steel (Figure 2.41)—shows a reasonably good overlap for AISI 304, whereas in mod 9Cr–1Mo steel, deviation from the experimental curve was observed in the membrane stretching regime. But when FE simulation was run for mod 9Cr–1Mo steel with an input data linearly extrapolated to larger strains, as shown in Figure 2.42, the point of plastic instability shifts to higher loads in FE simulation, resulting in a good overlap with experimental curve. This clearly shows that the input constitutive behavior plays a key role in the shape of the LDC.

To predict the tensile stress–strain curve from the SP curve, Manahan et al. (1986) developed a methodology using FE analyses, wherein the experimental LDC was iteratively matched with the database of curves generated by finite element analyses of the SP test.

Husain, Sehgal, and Pandey (2004) employed an inverse FE procedure to predict the modulus of elasticity and the true stress–strain curve. In the inverse finite element method, the experimental LDC is used as an input in a linear piecewise manner to the finite element program for comparison. The procedure involves an iterative piecewise matching of the initial slope (up to yielding) of the small punch experimental and FE-simulated load-displacement curves to generate the modulus value. Following this, a second

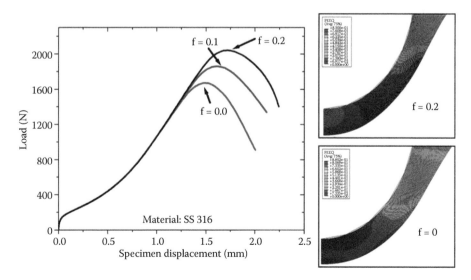

FIGURE 2.40
Effect of friction on LDC and plastic strain distribution in SP tests.

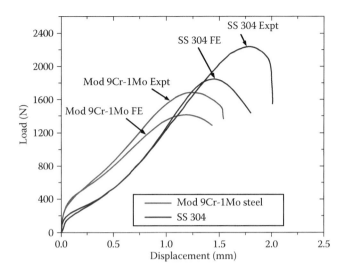

FIGURE 2.41
Comparison of the experimental LDC and FE-generated curve for SS 304 and mod 9Cr-1Mo steels.

linear section is examined iteratively using the first data point (σ_1, ep_1) of uniaxial true stress versus true plastic strain calculated in the previous step and assumed value of a second data point (σ_2, ep_2). The entire stress–strain curve is generated sequentially from the LDC using this approach. Comparison of the calculated values for three different steels generated iteratively with that of actual true stress–strain values shows encouraging results, although the uniqueness of the solution may not be exactly guaranteed.

In the identification procedure developed by Abendroth and Kuna (2003), a database of load-displacement curves is built using the finite element method via a systematic variation of material parameter, which is then used to train neural networks. The material properties are formulated as a constitutive damage law of Gurson, Tveergard, and Needleman (GTN) to simulate plasticity, damage, and fracture. The yield function in this model is defined as

$$\Phi(q, p, \sigma, f) = \left[\frac{q}{\sigma}\right]^2 + 2q_1 f^* \cosh\left[\frac{3}{2}q_2 \frac{p}{\sigma}\right] - [1 + q_3 f^{*2}] = 0$$

where
   σ is flow stress that related with the equivalent plastic strain
   f is the current void volume fraction
   $p = -\sigma_m$ with σ_m as the macroscopic mean stress and the von Mises effective
   stress $q = \sqrt{\dfrac{3}{2}S_{ij}S_{ij}}$
S_{ij} is the deviatoric components of the Cauchy stress tensor

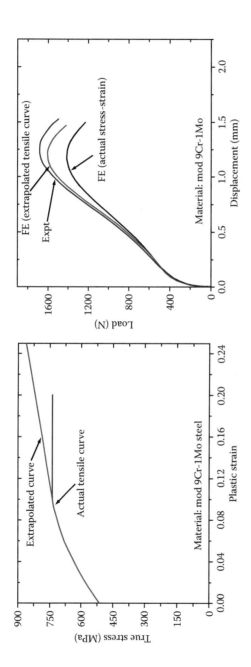

FIGURE 2.42

(a) Linearly extrapolated and actual true stress–strain curves of mod 9Cr-1Mo used in FE simulations. (b) Comparison of the experimental LDC of SP tests with FE-generated curves with extrapolated and actual stress–strain data inputs.

The constants q_1, q_2, and q_3 are fitting parameters, while f^* is the modified void volume fraction introduced by Tvergaard and Needleman (1984) to predict the rapid loss in strength that accompanies void coalescence and is expressed as

$$f^* = f \quad \text{if} \quad f \leq f_c$$

$$= f_c + \frac{f_u^* - f_c}{f_F - f_c}(f - f_c) \quad \text{if} \quad f \geq f_c$$

where f_c is the critical void volume fraction, f_F is the void volume fraction at final failure, which is usually $f_F = 0.15$, and $f_u^* = 1/q_1$ is the ultimate void volume fraction. The complete material characterization requires the determination of a number of parameters listed previously. Considering metallic materials, the number of parameters to be determined reduces to about seven: namely, σ_0, n, K, f_n, f_c, f_F, and ε_n, where the first three parameters are macromechanic ones and the rest are micromechnacal parameters for the damage model. The experimental LDC is transferred to a neural network (NN), which is trained using the LDC generated by FE simulations and the corresponding material parameters. During the training process, the NN generates an approximated function for the inverse problem relating the material parameters to the shape of LDC.

2.6 Concluding Remarks

Over the years, various specimen designs and methodologies have evolved for determination of tensile flow properties from small volumes of material. The design of test fixtures, customized equipment, and experimental design with respect to shear punch, small punch, and spherical indentation-based methods have achieved some state of uniformity in approach. These developments, along with the use of analytical tools and numerical methods such as finite element modeling and artificial neural networks in the analysis of small specimen test data for modeling the deformation behavior and for data inversion, have led to the present level of maturity. The next chapter dwells on the use of small specimens for fatigue- and fracture-toughness-related mechanical properties.

References

Abendroth, M., and Kuna, M. 2003. Determination of deformation and failure properties of ductile materials by means of the small punch test and neural networks. *Computational Materials Science* 28:633–644.

Ahn, J. H., and Kwon, D. 2001. Derivation of plastic stress–strain relationship from ball indentations: Examination of strain definition and pileup effect. *Journal of Materials Research* 16:3170–3178.

Armstrong, R. W. 1961. On size effects in polycrystal plasticity. *Journal of Mechanics and Physics of Solids* 9:196–199.

Atkins, A. G. 1980. On cropping and related processes. *International Journal of Mechanical Sciences* 22:215–231.

Au, P., Lucas, G. E., Scheckerd, J., and Odette, G. R. 1980. Flow property measurements from instrumented hardness tests. In *Non-destructive evaluation in the nuclear industry*, 579–610. New York: American Society for Metals.

Beghini, M., Bertini, L., and Fontanari, V. 2006. Evaluation of the stress–strain curve of metallic materials by spherical indentation. *International Journal of Solids and Structures* 43:2442–2459.

Beghini, M., Fontanari, V., and Monelli, B. D. 2009. Mechanical characterization of metallic materials by instrumented spherical indentation. In *Proceedings of the SEM Annual Conference*. Albuquerque, NM: Society for Experimental Mechanics Inc.

Biwa, S., and Storakers, B. 1995. An analysis of fully plastic Brinell indentation. *Journal of Mechanics and Physics of Solids* 43:1303–1333.

Butt, H. J., Cappella, B., and Kappl, M. 2005. Force measurements with the atomic force microscope: Technique, interpretation and applications. *Surface Science Reports* 59:1–152.

Byun, T. S., Kim, J. H., Chi, S. H., and Hong, J. H. 1998. Effect of specimen thickness on the tensile deformation properties of SA508 C1.3 reactor pressure vessel steel. In *Small specimen test techniques*, ASTM STP 1329, eds. W. R. Corwin, S. T. Rosinski, and E. V. Walle, 575–587. West Conshohocken, PA: ASTM.

Byun, T. S., Lee, E. H., Hun, J. D., Farrell, K., and Mansur, L. K. 2001. Characterization of plastic deformation in a disk bend test. *Journal of Nuclear Materials* 294:256–266.

Campitelli, E. N., Spatig, P., Bonade, R., Hoffelner, W., and Victoria, M. 2004. Assessment of the constitutive properties from small ball punch test: Experiment and modeling. *Journal of Nuclear Materials* 335:366–378.

Cao, Y., Qian, X., and Huber, N. 2007. Spherical indentation into elastoplastic materials: Indentation-response based definitions of the representative strain. *Materials Science and Engineering A* 454–455:1–13.

Cao, Y. P., and Lu, J. 2004. A new method to extract the plastic properties of metal materials from an instrumented spherical indentation loading curve. *Acta Materialia* 52:4023–4032.

Cheon, J. S., and Kim, I. S. 1996. Initial deformation during small punch testing. *Journal of Nuclear Materials* 24:255–262.

Demczyk, B. G., Wang, Y. M., Cumings, J., Hetman, M., Han, W., Zetti, A., and Ritchie, R. O. 2002. Direct mechanical measurement of tensile strength and elastic modulus of multiwalled carbon nanotubes. *Materials Science and Engineering* 334:173–178.

Dieter, G. E., Jr. 1961. *Mechanical metallurgy*. New York: McGraw–Hill.

Doerner, M. F., and Nix, W. D. 1986. A method for interpreting the data from depth sensing indentation instruments. *Journal of Materials Research* 1:601–609.

Field, J. S., and Swain, M. V. 1995. Determining the mechanical properties of small volumes of material from submicrometer spherical indentations. *Journal of Materials Research* 10(1):101–112.

Francis, H. A. 1976. Phenomenological analysis of plastic spherical indentation. *Transactions of ASME, Journal of Engineering and Materials Technology* 98(3):272–281.

George, R. A., Dinda, S., and Kasper, A. S. 1976. Estimating yield strength from hardness test. *Metals Progress* 30–35.

Gianola, D. S., and Eberl, C. 2009. Micro- and nano tensile testing of materials. *Journal of Minerals, Metals & Materials Society* (March):1–24. http://www.tms.org/pubs/journals/JOM/0903/gianola-0903.html.

Goyal, S., Karthik, V., Kasiviswanathan, K. V., Valsan, M., Bhanu Sankara Rao, K., and Raj, B. 2010. Finite element analysis of shear punch testing and experimental validation. *Materials Design* 31:2546–2552.

Guduru, R. K., Nagasekhar, A. V., Scattergood, R. O., Koch, C. C., and Murty, K. L. 2006. Finite element analysis of a shear punch test. *Metallurgical and Materials Transactions* 37A:1477–1483.

Guduru, R. K., Nagasekhar, A. V., Scattergood, R. O., Koch, C. C., and Murty, K. L. 2007. Thickness and clearance effects in shear punch testing. *Advanced Engineering Materials* 9:157–159.

Habbab, H., Mellor, B. G., and Syngellakis, S. 2006. Post-yield characterization of metals with significant pileup through spherical indentations. *Acta Materialia* 54:1965–1973.

Haggag, F. M. 1993. In situ measurements of mechanical properties using novel automated ball indentation system. In *Small specimen test techniques applied to nuclear reactor thermal annealing and plant life extension*, eds. W. R. Corwin, F. M. Haggag, and W. L. Server, 27–44. ASTM STP 1204. Philadelphia: American Society for Testing and Materials.

Haggag, F. M., Nanstad, R. K., Hutton, J. T., Thomas. D. L., and Swain, R. L. 1990. Use of automated ball indentation testing to measure flow properties and estimate fracture toughness in metallic materials. In *Applications of automation technology to fatigue and fracture testing*, eds. A. A. Braun, N. E. Ashbaugh, and F. M. Smith, 188–208. ASTM STP 1092. Philadelphia: American Society for Testing and Materials.

Hamilton, M. L., Toloczko, M. B., and Lucas, G. E. 1995. Recent progress in shear punch testing. In *Miniaturized specimens for testing of irradiated materials*, eds. H. Ullmaier and P. Jung, 46–59. IEA International Symposium.

Hankin, G. L., Toloczko, M. B., Hamilton, M. L., and Faulkner, R. G. 1998. Validation of the shear punch-tensile correlation technique using irradiated materials. *Journal of Nuclear Materials* 258–263:1657–1656.

Haque, M. A., and Saif, M. T. A. 2002. Mechanical behavior of 30–50 nm thick aluminum films under uniaxial tension. *Scripta Materialia* 47:863–867.

Hatanaka, N., Yamaguchi, K., and Takakura, N. 2003. Finite element simulation of the shearing mechanism in the blanking of sheet metal. *Journal of Materials Processing Technology* 139:64–70.

Hill, R., Storakers, B., and Zdunek, A. B. 1989. A theoretical study of the Brinell hardness test. *Proceedings of Royal Society London A* 423:301–330.

Huang, F. H., Hamilton, M. L., and Wire, G. L. 1982. Bend testing for miniature disk. *Nuclear Technology* 57:234–242.

Hurst, R., and Matocha, K. 2010. The European Code of Practice for Small Punch Testing—Where do we go from here? *Metallurgical Journal LXIII, Conference Proceedings of the 1st International Conference SSTT. Determination of Mechanical Properties of Materials by Small Punch and Other Miniature Testing Techniques,* 5–11, ISSN 0018-8069, ISBN 978-80-254-7994-0.

Husain, A., Sehgal, D. K., and Pandey, R. K. 2004. An inverse FE procedure for the determination for constitutive tensile behaviour of materials using miniature specimens. *Computational Materials Science* 31:84–92.

Igata, N., Miyaha, K., Ohno, K., and Uda, T. 1984. Proton irradiation creep of thin foil specimens of type 304 austenitic stainless steel and the thickness effects on their mechanical properties. *Journal of Nuclear Materials* 122 & 123:354–358.

Igata, N., Miyahara, K., Uda, T., and Asada, S. 1986. Effects of specimen thickness and grain size on the mechanical properties of types 304 and 316 austenitic stainless steel. In *The use of small-scale specimens for testing irradiated material,* ASTM STP 888, eds. W. R. Corwin and G. E. Lucas, 161–170. Philadelphia: ASTM.

Jaya Nagamani, B., and Alam, M. Z. 2013. Small scale mechanical testing of materials. *Current Science* 105:1073–1099.

Johnson, K. L. 1970. The correlation of indentation experiments. *Journal of Mechanics and Physics of Solids* 18:115–126.

Karthik, V., Laha, K., Kasiviswanathan, K. V., and Raj, B. 2002. Determination of mechanical property gradients in heat-affected zones of ferritic steel weldments by shear-punch tests. In *Small specimen test techniques,* ASTM STP 1418, eds. M. A. Sokolov, J. D. Landes, and G. E. Lucas, 380–406. West Conshohocken, PA: ASTM.

Karthik, V., Laha, K., Parameswaran, P., Chandravathi, K. S., Kasiviswanathan, K. V., Jayakumar, T., and Raj, B. 2011. Tensile properties of modified 9Cr–1Mo steel by shear-punch testing and correlation with microstructures. *International Journal of Pressure Vessels and Piping* 88:375–383.

Karthik, V., Visweswaran, P., Bhushan, A., Pawaskar, D., Kasiviswanathan, K. V., Jayakumar, T., and Raj, B. 2012. Finite element analysis of spherical indentation to study pileup/sink-in phenomena in steels and experimental validation. *International Journal Mechanical Sciences* 54:74–83.

Karthik, V., Visweswaran, P., Vijayraghavan, A., Kasiviswanathan, K. V., and Raj, B. 2009. Tensile–shear correlations obtained from shear punch test technique using a modified experimental approach. *Journal of Nuclear Materials* 393:425–432.

Kasiviswanathan, K. V., Hotta, S. K., Mukhopadyay, C. K., and Raj, B. 1998. Miniature shear punch test with online acoustic emission monitoring for assessment of mechanical properties. In *Small specimen test techniques,* ASTM STP 1329, eds. W. R. Corwin, S. T. Rosinski, and E. V. Walle. Philadelphia: ASTM.

Kiener, D., Grosinger, W., Dehm, G., and Pippan, R. 2008. A further step towards an understanding of size-dependent crystal plasticity: In situ tension experiments of miniaturized single-crystal copper samples. *Acta Materialia* 56:580–592.

Kim, J. H., and Kim, S. H. 2013. Microstructure and mechanical property of ferritic–martensitic steel cladding under a 650°C liquid sodium environment. *Journal of Nuclear Materials* 443:112–119.

Klingenberg, W., and Singh, U. P. 2003. Finite element simulation of the punching/blanking process using in-process characterization of mild steel. *Journal of Materials Processing Technology* 134:296–302.

Klingenberg, W., and Singh, U. P. 2005. Comparison of two analytical models of blanking and proposal of a new model. *International Journal of Machine Tools & Manufacture* 45:519–527.

Konopik, P., and Dzugan, J. 2012. Determination of tensile properties of low carbon steel and alloyed steel 34CrNiMo6 by small punch test and microtensile test. In *Determination of mechanical properties of materials by small punch and other miniature testing techniques*, eds. K. Matocha, R. Hurst, and W. Sun, 319–328. Conference Proceedings: The 1st International Conference SSTT: August 31 to September 2, 2010. Ostrava, Czech Republic.

LaVan, A., and Sharpe, W. N. 1999. Tensile testing of microsamples. *Experimental Mechanics* 39:210–216.

Lee, H., Lee, J. H., and Pharr, G. M. 2005. A numerical approach to spherical indentation techniques for material property evaluation. *Journal of Mechanics and Physics of Solids* 53:2037–2069.

Lee, J. H., Kim, T., and Lee, H. 2010. A study of robust indentation techniques to evaluate elastic–plastic properties of metals. *International Journal of Solids and Structures* 47:647–664.

Lord, J. D., Roebuck, B., Morrell, R., and Lube, T. 2010. Aspects of strain and strength measurement in miniaturized testing for engineering metals and ceramics. *Materials Science Technology* 26:127–148.

Lucas, G. E. 1983. The development of small specimen mechanical test techniques. *Journal of Nuclear Materials* 117:327–339.

Lucas, G. E. 1990. Review of small specimen test techniques for irradiation testing. *Metallurgical and Materials Transactions A* 21A:1105–1119.

Lucas, G. E., Sheckherd, J. W., and Odette, G. R. 1986. Shear punch and microhardness tests for strength and ductility measurements. In *The use of small scale specimens for testing irradiated material*, ASTM STP 888, eds. W. R. Corwin and G. E. Lucas, 112–140. Philadelphia: ASTM.

Lucas, G. E., Sheckherd, J. W., Odette, G. R., and Panchanadeeswaran, S. 1984. Shear punch tests for mechanical property measurements in TEM disc-sized specimens. *Journal of Nuclear Materials* 122:429–434.

Manahan, M. P., Argon, A. S., and Harling, O. K. 1981. The development of a miniaturized disk bend test for the determination of post-irradiation mechanical properties. *Journal of Nuclear Materials* 104:1545–1550.

Manahan, M. P., Browning, A. E., Argon A. S., and Harling O. K. 1986. Miniaturized disc bend test technique development and application. In *The use of small scale specimens for testing irradiated specimens*, STP 888, eds. W. R. Corwin and G. E. Lucas, 17–49. Philadelphia: ASTM.

Mao, X., and Takahashi, H. 1987. Development of a further-miniaturized specimen of 3 mm diameter for tem disk (ø 3 mm) small punch tests. *Journal of Nuclear Materials* 150:42–52.

Matocha, K. 2015. Small-punch testing for tensile and fracture behavior: Experiences and way forward in small specimen test techniques, vol. 6, ASTM STP 1576, eds. M. A. Sokolov and E. Lucon, 145–159. West Conshohocken, PA: ASTM.

Matthews, J. R. 1980. Indentation hardness and hot pressing. *Acta Metallurgica* 28:311–318.

Mesarovic, S. D., and Fleck, N. A. 1999. Spherical indentation of elastic–plastic solids. *Proceedings of Royal Society London A* 455:2707–2728.

Miyazaki, S., Shibata, K., and Fujita, H. 1979. Effect of specimen thickness on mechanical properties of polycrystalline aggregates with various grain sizes. *Acta Metallurgica* 27:855–862.

Norbury, A. L., and Samuel, T. 1928. The recovery and sinking-in or piling-up of material in the Brinell test and the effects of these factors on the correlation of the Brinell with certain other hardness tests. *Journal of Iron and Steel Institute* 117:673–675.

Okada, A., Lucas, G. E., and Kiritani, M. 1988. Micro-bulge test and its application to neutron-irradiated metals. *Transactions of Japan Institute of Metals* 29 (2):99–108.

Oliver, W. C., and Pharr, G. M. 1992. An improved technique for determining hardness and elastic modulus using load and displacement sensing indentation techniques. *Journal of Materials Research* 7:1564–1583.

Partheepan, G., Sehgal, D. K., and Pandey, R. K. 2006. An inverse finite element algorithm to identify constitutive properties using dumb-bell miniature specimen. *Modeling Simulation Materials Science and Engineering* 14:1433–1445.

Ramaekars, J. A. H., and Kals, J. A. G. 1986. Strains, stresses and forces in blanking. In *Proceedings of the IMC Conference*, Galway, 126–138.

Sharma, K., Bhasin, V., Vaze, K. K., and Ghosh, A. K. 2011. Numerical simulation with finite element and artificial neural network of ball indentation for mechanical property estimation. *Sadhana* 36:181–192.

Sharpe, W. N., Danley, D., and LaVan, A. 1998. Microspecimen tensile tests of A533-B steel. In *Small specimen test techniques*, ASTM STP 1329, eds. W. R. Corwin, S. T. Rosinski, and E. V. Walle, 497–512. West Conshohocken, PA: ASTM.

Smith, D. A. 1990. *Die design handbook*, 3rd ed., 46. Dearborn, MI: Society of Manufacturing Engineers.

Sneddon, I. N. 1965. The relation between load and penetration in the axisymmetric Boussinesq problem for a punch of arbitrary profile. *International Journal of Engineering Science* 3:47–57.

Tabor, D. 1951. *The hardness of metals*. Oxford, England: Clarendon Press.

Taljat, B., and Pharr, G. M. 2004. Development of pileup during spherical indentation of elastic–plastic solids. *International Journal of Solids and Structures* 41:3891–3904.

Taljat, B., Zacharia, T., and Kosel, F. 1998. New analytical procedure to determine stress–strain curve from spherical indentation data. *International Journal of Solids and Structures* 35:4411–4426.

Tirupataiah, Y., and Sundararajan, G. 1991. On the constraint factor associated with the indentation of work-hardening materials with a spherical ball. *Metallurgical and Materials Transactions* 22A:2375–2384.

Toloczko, M. B., Abe, K., Hamilton, M. L., Garner, F. A., and Kurtz, R. J. 2002a. The effect of test machine compliance on measured shear punch yield stress as predicted using finite element analysis. In *Small specimen test techniques*, ASTM STP 1418, eds. M. A. Sokolov, J. D. Landes, and G. E. Lucas, 339–349. West Conshohocken, PA: ASTM International.

Toloczko, M. B., Hamilton, M. L., and Lucas, G. E. 2000. Ductility correlations between shear punch and uniaxial tensile test data. *Journal of Nuclear Materials* 283:987–991.

Toloczko, M. B., Kurtz, R. J., Hasegawa, A., and Abe, K. 2002b. Shear punch tests performed using a new low compliance test fixture. *Journal of Nuclear Materials* 307–311:1619–1623.

Toloczko, M. B., Yokokura, Y., Abe, K., Hamilton, M. L., Garner, F. A., and Kurtz, R. J. 2002c. The effect of specimen thickness and grain size on mechanical properties obtained from the shear punch test. In *Small specimen test techniques,* ASTM STP 1418, eds. M. A. Sokolov, J. D. Landes, and G. E. Lucas, 371–379. Philadelphia: ASTM.

Tvergaard, V., and Needleman, A. 1984. Analysis of the cup-cone fracture in a round tensile bar. *Acta Metallurgica* 32:157–169.

Zhao, M., Ogasawara, N., Chiba, N., and Chen, X. 2006. A new approach to measure the elastic–plastic properties of bulk materials using spherical indentation. *Acta Materialia* 54:23–32.

Zhou., H., Henson, B., Bell, A., Blackwood, A., Beck, A., and Burn, R. 2006. Linear piezo-actuator and its applications. http://zhouhx.tripod.com/piezopaper.pdf.

Zhou, Q., and Wierzbicki, T. 1996. A tension model of blanking and tearing of ductile metal plates. *International Journal Mechanical Sciences* 38:303–324.

3

Miniature Specimens for Fatigue and Fracture Properties

3.1 Subsize Charpy Specimen Impact Testing

Charpy V-notch (CVN) impact testing, which employs a 10 mm square bar of 55 mm length, is one of the test methods widely used to evaluate the toughness characteristics of a material. Using an instrumented pendulum hammer and precracked and side-grooved specimen, force-time, force-displacement, and energy-displacement curves can be generated for evaluating fracture toughness data in terms of J or K_I.

A variety of scaled-down versions of conventional CVN specimens (Louden et al. 1988; Lucas et al. 1988; Lucon 1998) have been investigated, including the half-size ($5 \times 5 \times 23.6$ mm^3 long) and third-size ($3.33 \times 3.33 \times 23.6$ mm^3 long) specimens as shown in Figure 3.1, as well as various other sizes down to dimensions as small as $1 \times 1 \times 20$ mm^3 (Kayano et al. 1991). The primary motivation for the use of subsize CVN specimens was the need for monitoring the embrittlement of reactor pressure vessel (RPV) steels of light water reactors (LWRs). The utilization of subsize specimens greatly benefited the materials surveillance programs for life extension of LWR pressure vessels using the available RPV archive materials. However, the main issue has been scaling of parameters such as absorbed energy and ductile-brittle transition temperature (DBTT).

The energy absorbed in fracturing a notched specimen is a complex function of both the elastic and plastic deformation in the specimen prior to and during crack initiation and propagation from the notch root. These processes are sensitive to stress state and hence the specimen size. Decreasing the specimen size minimizes the constraints for plain strain conditions (necessary for brittle failure) as compared to full size (FS) specimens and promotes ductile failure at a given temperature, thereby decreasing the DBTTs. The absorbed energy is reduced when specimen size is reduced due to the smaller volume involved in the fracture process. The earliest attempts to correlate upper shelf energy (USE) and DBTT from subsize specimens with that of FS specimens were based on normalization factors (NFs). Common NFs

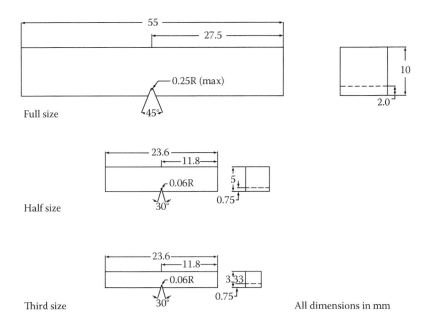

FIGURE 3.1
Schematic showing the full size and subsize CVN specimen geometries.

include fracture area, fracture volume, stress concentration at the notch root, or a combination of these factors.

3.1.1 Energy Correlation

Most of the early studies (Corwin and Hougland 1986; Lucas et al. 1988; Kumar, Garner, and Hamilton 1990; Lucon 1998) revealed that parameter fracture volume (Bb^2, where B = specimen thickness and, b = specimen ligament below notch root) worked well for normalizing USE of subsize and FS specimens of ductile materials (USE > 200 J), while for materials in brittle condition with predominantly flat fractures, the factor B.b (fracture area) yielded better correlation. Corwin, Klueh, and Vitek (1984) employed a slightly different fracture volume factor ($Bb^{3/2}$), which worked well for ductile materials of USE > 150 J. Normalization factors taking into account both specimen dimensions and notch geometry such as Bb^2/LK_t (where L is the specimen span and K_t is the stress concentration factor as a function of notch root radius and specimen dimensions) were shown to accurately normalize the USE over a range of USE from 300 to below 100 J (Louden et al. 1988).

To develop normalization factors applicable to both ductile and brittle conditions, Kumar et al. (1993) partitioned USE into two components: namely, those required for crack initiation and for crack propagation. While the USE of notched specimens is the sum of previously mentioned components, USE of precracked specimens reflects only crack propagation component (USE_p).

The difference in USE values between the two cases (USE = USE – USE_p) represents the macrocrack initiation energy expended in deforming the material in the fracture volume. USE correlated well with the fracture volume (Bb^2); that is,

$$\left(\frac{\Delta USE}{\text{Fracture volume}} \right)_{\text{full size}} = \left(\frac{\Delta USE}{\text{Fracture volume}} \right)_{\text{subsize}} \tag{3.1}$$

The ratio of USE and USE_p was also shown to be invariant with specimen size. Using this partition methodology involving both notched and precracked specimens, USE predicted for FS specimens by Kumar et al. (1993) was found to be within ±7% of the actual value, particularly in the lower USE range of 50–200 J.

Kayono et al. (1991) considered the effects of notch angle and plastic constraint in the methodology for USE correlation and included Q, the plastic stress concentration factor ($Q = 1 - \frac{\theta}{2} - \frac{\pi}{2}$, where θ is notch angle) in the fracture volume factor. Incorporating volume factor as $(Bb^2 Q_{FS})/(K_t Q_{subsize})$, the normalized USE of subsize CVN specimens as small as $1 \times 1 \times 20$ mm^3 and $1.5 \times 1.5 \times 20$ mm^3 correlated well with that of FS specimens. This methodology was further improved (Schubert et al. 1995) for USE prediction by including the specimen span (L) in the normalization factor as (Bb^2/K_tQL). The predicted values of FS specimen USE of A533B steel in both irradiated and unirradiated condition, based on half-size ($5 \times 5 \times 23.6$ mm^3) and third-size ($3.33 \times 3.33 \times 23.6$ mm^3), were within 10% of data for FS specimens.

In another exhaustive work undertaken by Sokolov and Nanstad (1995), with different subsize CVN specimen geometry (5×5, 3×4, 3.3×3.33 mm^2) with different notch angles (30°, 45°, 60°) and spans (L = 20, 22, 40 mm) on four RPV steels, the effects of specimen dimensions, including depth, angle, and radius of notch, were studied. Increasing the depth of notch significantly reduced the USE. Varying notch angle from 30° to 45° while keeping the remaining dimensions identical did not affect USE, and span as well as impact velocity (in the range of 2.25 to 5.5 m/s) did not affect the USE and DBTT. Observing that no single known correlation procedure would work for data from different subsize specimens, Sokolov and Nanstad formulated a new correlation methodology for absorbed energy by partitioning the fracture process into low-energy brittle and high-energy ductile modes as

$$E = E_{subsize} * \left[NF_{brittle} * \frac{100 - SHEAR}{100} + NF_{ductile} \frac{SHEAR}{100} \right] \tag{3.2}$$

where $(NF)_{brittle} = (Bb)_{FS}/(Bb)_{subsize}$, $NF_{ductile}$ was geometry-specific empirical factors, and SHEAR the percentage of shear fracture on the fracture surface.

3.1.2 Transition Temperature Correlation

There are numerous definitions for ductile–brittle transition temperature. The temperature indexed to a particular value of absorbed energy, lateral expansion, or fracture appearance is taken as DBTT. It is commonly defined as the temperature at the midpoint between the upper and lower shelf energies. As regards the DBTT correlation of FS and subsize specimens, the methodologies developed are based on the use of either empirical relationships relating DBTT to specimen size/geometry or use of a normalization factor such as critical cleavage fracture stress. Based on the hypothesis that fracture is controlled by maximum tensile stress ahead of a crack tip normal to the crack plane, Kumar et al. (1993) proposed that the crack propagates, causing fracture when the stress exceeds a critical value of σ'. A normalized value of DBTT was defined as the ratio of DBTT and σ' where σ' is the maximum elastic tensile stress at the notch root expressed as

$$\sigma' = \frac{K_t L 3 P_m}{2Bb^2} \tag{3.3}$$

where P_m is the maximum load observed in a Charpy test at the point of general yielding, K_t is the stress concentration factor, and L is the specimen span. Computing P_m in terms of the cleavage fracture stress (σ_f), it was shown using data of various steels that normalized DBTT ($DBTT_n = DBTT/\sigma'$) for FS and subsize specimens could be related as

$$(DBTT_n)_{FS} = (DBTT_n)_{subsize} + constant \tag{3.4}$$

where the constant was both size and material dependent, but independent of alloy condition (precracking or thermal aging). The shift in DBTT due to heat treatment or precracking (Kumar et al. 1993) was found to be equal for FS and subsize specimens

$$\Delta(DBTT_n)_{FS} = \Delta(DBTT_n)_{subsize} \tag{3.5}$$

This equality was also seen to be valid for DBTT change measured on FS, half-size ($5 \times 5 \times 23.6$ mm^3) and third-size ($3.33 \times 3.33 \times 23.6$ mm^3) specimens of A533B steel due to neutron irradiation (Schubert et al. 1995).

In a different approach, Kurishita et al. (2004) formulated DBTT correlation based on the premise that specimen size effects on DBTT arise due to constraint loss. A constraint factor, α that is an index of the plastic constraint was defined as $\alpha = \sigma^*/\sigma_y$, where σ^* is the critical cleavage fracture stress and a material constant and σ_y is the uniaxial yield stress at the DBTT at the strain rate generated in the Charpy impact test. Using finite element simulations of the impact test and a cleavage fracture model that is based on a critical stress–critical area plot, the critical cleavage stress σ^* and the stressed area A^* were computed. The parameter α was seen to correlate with the parameter A^*/b^2 for differently sized CVN samples of a ferritic martensitic steel.

Other studies have resulted in deriving material-specific empirical factors between FS and subsize specimens. Korolev et al. (1998) compared the values of DBTT obtained for subsize ($5 \times 5 \times 27$ mm³) and FS Charpy specimens with identical notch configuration for pressure vessel material of the Russian water–water energetic reactor (VVER) types of reactors in its irradiated condition. The criterion for definition of DBTT using subsize specimens was based on the law of similar deformation in solids, where the constancy of ratio of the level of energy to the USE is ensured. Establishing that the criterion level of absorbed energy 47 J in standard Charpy specimens corresponds to 6 J in 5×5 mm² specimens, the following correlation was recommended:

$$DBTT^{10-10} = 50 + DBTT^{5-5} \pm 2\sigma, \,°C \quad (3.6)$$

where σ is 21°C.

Similarly, Bohme and Schmitt (1998) established empirical relations between the FS ISO Charpy specimens of nuclear pressure vessel steels and the subsize version designated as KLST (Kleinst–Proben) with dimensions of $27 \times 4 \times 3$ mm³, 1.0 mm notch depth, and 0.1 mm radius as

$$T_{KLST} = T_{ISO} - 55°C \quad (3.7)$$

The transition temperatures in KLST corresponded to energies of 1.9 and 3.1 J, while those in standard CVN tests corresponded to levels of 41 and 68 J.

Recently, Moitra et al. (2015) derived a power law relationship for the DBTT of small-size specimens to that of FS specimens based on the ratio of notch root volumes of subsize ($V_{Nsubsize}$) and FS (V_{NFS}) specimens as

$$\frac{DBTT_{FS}}{DBTT_{subsize}} = \left(\frac{V_{NFS}}{V_{subsize}} \right)^m \quad (3.8)$$

where m varies exponentially from 0.05 to 0.25 with respect to percentage of USE (20%–80%).

Using finite element (FE) simulations, it was further inferred for various subsize geometries of mod 9Cr-1Mo steel that the total stress at the mid-section along the crack propagation direction for different specimen sizes reaches the cleavage fracture stress (2400–2450 MPa based on FS CVN) at temperatures corresponding to 33% of respective USE. It may be noted that the 33% USE corresponds to ~68 J of FS specimens used in the determination of nil ductility transition temperature.

3.1.3 Scaling of Instrumented Impact Test Parameters

Toward predicting toughness-related properties from instrumented tests using subsize specimens, Schindler and Veidt (1998) formulated scaling laws for the key parameters of the load-displacement curve (Figure 3.2) such as maximum force (F_m), energy at maximum force (W_{mp}), and total fracture energy (W_{tot}). Carrying the analysis for a typical upper shelf behavior of ductile tearing consisting of a crack tip blunting and crack initiation phase (J-integral controlled) and a tearing crack growth phase governed by crack tip opening angle (CTOA), the scaling equations for total energy (W_{tot}) and maximum force (F_m) were derived as

$$W_{tot} = \frac{B}{B'} \frac{\eta'}{\eta} \left[\left(\frac{b_0}{b_0'} \right)^{p+1} W_{mp}' + \left(\frac{b_0}{b_0'} \right)^2 (W_{tot}' - W_{mp}') \right] \tag{3.9}$$

$$F_m = \frac{B}{B'} \left(\frac{b_0}{b_0'} \right)^2 \frac{S'}{S} F_m' \tag{3.10}$$

where W, B, S, b_0, and a_0 are geometrical parameters of full size specimens, while those of subsize specimens are designated by a prime ('). η is factor f (a/W) for three-point bending loading. For two types of subsize specimens—namely, $3 \times 4 \times 27$ and $5 \times 5 \times 27$ mm^3, the W_{tot} and F_m, estimated through theoretically derived scaling laws given by Equations 3.9 and 3.10, were seen to agree with experimental values of FS specimens within ±5%.

3.1.4 Challenges in Subsize Impact Testing

In addition to the challenges in data interpretation from subsized specimens, the small CVN specimens introduce experimental constraints such as rapid

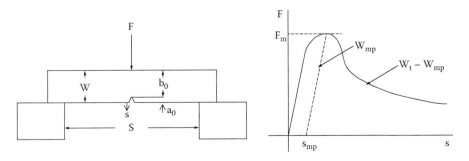

FIGURE 3.2
Typical load-displacement curve of an instrumented impact test.

temperature change and inertial vibrations. The temperature loss in subsize specimens during its transfer from the furnace to the anvil support until the impact is overcome by using fast automated transfer mechanisms or in situ heating/cooling arrangements (Manahan 2000), ensuring close temperature control.

Because of low stiffness, subsize Charpy specimens vibrate at higher frequency than FS specimens. Electronic measuring and recording devices of higher upper frequency bounds (about 250 kHz) are required for correct registration of the high-frequency oscillations. Instrumented tups of high sensitivity are required for accurate recording of small loads in subsize impact tests. The inertial vibrations, which can influence the yield load determination, can be overcome by decreasing the impact velocity and increasing the mass of the striker. However, the advantages of subsize CVN specimens in terms of reduced volume override the difficulties associated with experiments and data interpretation.

3.2 Fracture Toughness K_{IC}/J_{IC} with Subsize Specimens

Fracture toughness is dependent on specimen size as the crack tip stress and strain fields are influenced by specimen dimensions. The recommended specimens for the plane strain fracture toughness K_{IC} testing are the three-point bend SE(B) specimen or the compact tension (CT) specimen. The determination of K_{IC} per ASTM E 399 requires thick specimens to establish plain strain condition at the crack tip specified as

$$B, a \geq 2.5 \left(\frac{K_{IC}}{\sigma_{ys}} \right)^2 \tag{3.11}$$

where B is specimen thickness, a is crack length, and σ_{ys} is the yield strength. For a typical ferritic steel with $K_{IC} \sim 100$ MPa$\sqrt{m}$ and $\sigma_{ys} \sim 400$ MPa, the specimen thickness must be >150 mm for a valid test. For all practical purposes, the size requirements preclude the use of subsize specimens.

For fracture toughness determined using the J-integral procedure, which takes into account the nonlinear material behavior, the size requirements are more lenient, such as

$$B, b \geq 25 \frac{J_{IC}}{\sigma_{ys}} \tag{3.12}$$

where b is the uncracked ligament and J_{IC} is the resistance to initiation of stable crack growth. The J integral procedure enables testing with samples thinner than the ASTM E 399 specimens by a factor of 20. In the case of neutron-irradiated steels, because of irradiation hardening and postirradiation toughness degradation, the plastic zone near the crack tip and hence the minimum thickness required for a valid J_{IC} further reduce.

Huang (1988) has carried out extensive work on use of round CT specimens of thickness from 12 down to 1.02 mm for evaluating J_{IC} in both unirradiated and irradiated austenitic and ferritic steels (Figure 3.3). The study showed that, in addition to size criteria, restriction on crack extension for ensuring J-controlled crack growth is important for small specimens. The thick independence of J_{IC} was shown for 11.94, 7.62, and 2.54 mm thick CT specimens of HT9 alloy tested at 28°C and 232°C. However, a measurement uncertainty of 15% in J_{IC} values was reported. Using specimen thickness ranging from 25 to 1 mm, Mao (1991) formulated a modified J_{IC} procedure for invalid specimen size by shear fracture measurements using recrystallization etch method and fractographic observations and related the J-integral of mixed mode fracture with that of plain strain condition using rigid plastic analysis.

Specimen size effects were further studied by Ono, Kasada, and Kimura (2004) by employing two size variants: namely, reduced thickness with other dimensions being the same as a standard (1 CT) specimen (i.e., 1/2 T – 1 CT) and miniaturized CT with proportionally reduced dimensions (1/2 CT, 1/4 CT). The fracture toughness, J_Q, increased as the specimen thickness decreased, while J_Q decreased when the specimens were miniaturized maintaining similar proportions. With decrease in specimen thickness, the plane stress state becomes predominant and the plastic zone size at the crack tip increases near specimen side surfaces. Since the energy spent on the plastic deformation increased, the fracture toughness increased. However, in the latter case of proportionally minimized specimen dimensions such as 1/2 CT and 1/4 CT, the plastic zone size becomes too small to suppress gross yielding, resulting in flow instability and a decrease in the J_Q values for the smaller specimens.

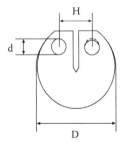

		Specimen type		
Dimension	A	B	C
D	16	18	20
d	3.2	3.4	4
H	6.5	7.3	8.14
B(thk)	5.92	6.67	7.4

FIGURE 3.3
Round compact tension specimen geometry with different possible combinations of dimensions for fracture toughness determination. (From Huang, F. 1986, Use of subsized specimens for evaluating the fracture toughness of irradiated material. *In The use of small-scale specimens for testing irradiated material*, ASTM STP 888, eds. W.R. Corwin and G.E. Lucas, 290–304. Philadelphia: American Society for Testing and Materials.)

3.2.1 Toughness in Transition Regime

The evaluation of the fracture toughness of pressure vessel steels in the transition regime was earlier established through correlations based on the Charpy transition curve. Though the shape of the toughness versus temperature curve in the transition regime is virtually similar for all ferritic steels, the scatter in the toughness data in the transition region was quite large necessitating a large number of tests for statistical analysis. Based on the weakest link theory and the Weibull statistics theory, Wallin (1999) developed a master curve method with a corresponding reference temperature to provide a description of the statistical scatter, statistical size effects, and temperature dependence of cleavage toughness in the transition region. In the master curve approach, the fracture toughness in the transition regime is described with only one parameter: the reference temperature T_0, defined as the temperature at which the mean fracture toughness for a 1 in. (25.4 mm) thick fracture toughness specimen equals 100 MPa $\sqrt{m}$. It is the main parameter to define the curves for the mean and lower bound fracture toughness. The schematic presentation of master curve size adjustment is shown in Figure 3.4.

The fracture toughness master curve allows the fracture toughness corresponding to a particular temperature or section thickness to be estimated from data generated at a different temperature and/or section thickness, by referencing to an indexing temperature, T_0. A comparison made by Wallin et al. (2001) between T_0 and scatter estimates from test results measured with miniature specimens (i.e., smaller than the Charpy size) showed that the definitions of scatter and the measuring capacity (specimen size) criterion apply even to miniature specimens that are of 3×4 mm^2 and 3.3×3.3 mm^2 cross section SE(B). The comparison demonstrated, in all respects, applicability of the miniature size specimens for the fracture toughness estimation using the master curve approach.

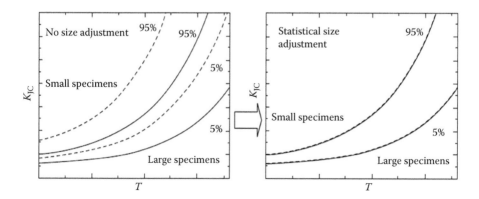

FIGURE 3.4
Master curve methodology.

The small specimen fracture toughness data in the transition regime are influenced by the lack of constraint with the possibility of deviating from small-scale yielding. Odette and co-workers (1998) developed the master curve shift (MC-ΔT) method using small specimens accounting for the effects of geometry, irradiation, loading rates, and safety margins. They demonstrated the use of a relatively fewer number of miniature specimens to index a baseline K_e, where the curve is adjusted in temperature space by an amount ΔT_g to conditions of higher constraint associated with specimens of larger dimensions. The position of the curve, as shown in Figure 3.5, is adjusted using additional shifts, ΔT, to account for strain rate, irradiation, weakest link types of size effects, and safety margins. Microstructure-based models have been developed to predict these shifts and the shape and position of the reference curve.

Systematic studies were undertaken by Odette, at UCSB in conjunction with NRG, Petten and CRPP, Switzerland to address size effects on effective fracture toughness $K_e(T)$ due to weakest link statistics and constraint effects. A statistical size adjustment (SA) scaling with $B^{1/4}$, where B is the fracture specimen thickness, was derived from weakest link statistics per the ASTM E1921 procedure. The constraint loss adjustment (CLA) based on both B (specimen thickness) and fracture specimen ligament size, b, was derived by combining finite element method (FEM) simulations of crack tip fields with critical stress (σ^*)–critical area (A^*) and micromechanics of cleavage fracture in body-centered

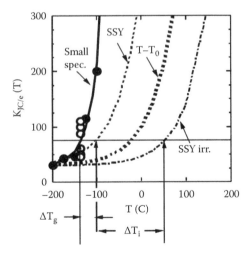

FIGURE 3.5
Schematic illustration of the MC-ΔT method. The heavy dashed curve shows the basic MC shape on a T–T_0 reference scale. The solid curve represents a small specimen MC with a steeper slope and lower T_0 (referenced at 75 MPa$\sqrt{m}$) by ΔT_g than the corresponding small scale yielding (SSY) curve dotted line. It has been temperature indexed with a small number of small specimens. Irradiation shifts the SSY curve by an additional increment ΔT_i, shown by the dashed-dotted line. (From Lucas, G.E. et al., *Journal of Nuclear Materials* 307–311:1600–1608, 2002.)

cubic (bcc) metals (Lucas et al. 2007). This approach was based on the analysis of fracture surface using confocal microscopy and fracture reconstruction techniques (Lucas et al. 2007). The same research group at UCSB developed a small precracked bend bar called the deformation and fracture minibeam (DFMB), nominally one-sixth in size in all dimensions compared to a standard CVN specimen for measuring shifts in the cleavage transition temperature under dynamic loading. The shape of the $K_e(T)$ for the DFMB specimen was found to have steeper transition compared to the MC shape found with larger specimens. Considering the advantages of significant savings in the reactor space, the DFMB specimens have been extensively employed for determining (ΔT_d) of neutron irradiated steels and compared with ΔT_i shifts determined previously from tests on 1/3-sized pre-cracked Charpy bend bars.

3.3 Specimen Reconstitution Methods

Specimen reconstitution involves preparing standard specimen geometry by compounding a small test piece with undeformed portions of a previously tested specimen and employing them for conventional tests. It is a powerful tool, especially for toughness evaluation using Charpy impact and CT specimens, where size requirements for a valid test demand large volumes of material. The general method for a reconstituted Charpy specimen is to machine the failed Charpy portions and weld them with the test material at the center (Figure 3.6). The completed blank after welding is machined into another Charpy specimen (Viehrig and Boehmert 1988). Different configurations of reconstituted CT specimens (Tomimatsu, Kawaguchi, and Iida 1998) obtained from the halves of Charpy specimens are shown in Figure 3.7. It is essential to ensure that the loaded insert material is not influenced by the welding and machining procedure. To this end, various welding approaches, such as laser welding, electron beam welding, flash welding, arc stud welding, and friction welding, have been investigated. The main consideration of the choice of welding process is the small size of the heat-affected zone (HAZ); low heat input, which minimizes the possibility of microstructural changes in the insert material; and the residual stress field, which could influence the crack growth. The successful use of a reconstituted CT specimen is also dependent on the possible interference between the plastic zone and the weld.

Toward developing a set of recommended practices for reconstitution of nonirradiated and irradiated Charpy size specimens, a project coded RESQUE, initiated among European institutions during 1998–2000, has formulated guidelines in all critical parameters in reconstitution technique. These include recommendations for preparation of materials and reference testing, heat input, qualification of weld, lowest acceptable insert dimensions, temperature measurements, verification for irradiated material, etc. (van Walle et al. 2002).

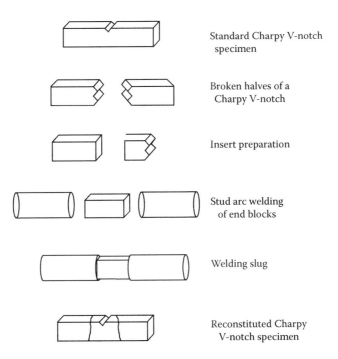

FIGURE 3.6
Reconstitution of Charpy impact specimen from broken halves.

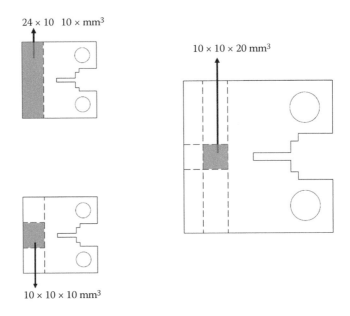

FIGURE 3.7
CT specimen geometry reconstituted from broken halves of Charpy samples.

3.4 Fatigue and Fatigue Crack Growth Studies Using Subsize Specimens

In fatigue tests, it is known that fracture is governed by internal defects in the low cycle region, while cracking occurs on the specimen surface and propagates inward in the high cycle region. Therefore, fatigue properties depend on the specimen dimension. Specimen size effect has been reported when the number of grains at the minimum cross section in the gauge section is smaller than a critical value that ranges from 5 to 10 depending on the tests and materials. Liu and Grossbeck (1986) were the first to develop fatigue test techniques using subsize specimens for test temperatures up to 650°C. They demonstrated that with half the size of an ASTM-606 specimen (i.e., 3.18 mm diameter gauge section), the S–N (stress amplitude versus number of cycles to failure) data were similar to that from large specimens for both unirradiated and irradiated SS316 steel. Though subsize specimens permitted the use of fully reversed cyclic fatigue loading to a total strain range of 2%, fatigue life fell consistently below that of FS specimens.

Hourglass types of specimens have good resistance to buckling, which is a very important issue for miniaturizing specimens for push–pull tests. Hirose et al. (2001) demonstrated the use of subsize hourglass specimens of 2 mm gauge diameter and 7 mm in length to achieve high-strain ranges in push–pull fatigue tests and good temperature control. On the other hand, there are some difficulties in estimating the axial strain range from the results of the diametric strain of hourglass specimens. For this reason, the choice of a specimen configuration with a cylindrical gauge section has been recommended by several researchers. Using FE simulations, Moslang (2000) has developed miniaturized cylindrical specimen geometry with a gauge diameter of 2 and 7.6 mm in length and shown S–N results and cyclic stress–strain curves in good agreement with standard specimen tests.

Pahlavanyali et al. (2008) attempted to use miniature specimens with a gauge length of 14 mm and a cross section of about 1 mm^2 for thermomechanical fatigue (TMF) behavior of Nimonic 90. The results from miniature specimens showed that stress evolution and failure mechanism were similar to the conventional test piece, but the TMF lives were shorter by between 20% and 30%.

The limited database on small specimen fatigue testing is due to the complexity of test conditions such as imposing push–pull loading and tension–tension loading, the cost involved, and the difficulty in interpreting the test data due to specimen size effects.

3.4.1 Fatigue Crack Growth Studies

Standard fatigue crack propagation test places strict size requirement on the specimen to ensure that net section yielding is minimal and the crack tip stress field can be correctly represented by the linear elastic fracture mechanics parameter ΔK. In a small specimen, this size requirement may be violated at some point of the test and crack growth data may deviate from the standard test results.

The earliest study on fatigue crack growth (FCG) by Ermi and James (1986) has shown that specimen thickness is not a major constraint for measuring subcritical crack growth. This was evident from the good agreement of fatigue crack growth studies on miniaturized center cracked tension (CCT) specimens with thicknesses of 0.254 and 0.457 mm (planar dimensions of 50 mm wide, 203 mm long) with results of specimens from thicker plate material. This provided the confidence to reduce the specimen sizes to extremely small sizes for valid crack growth studies. Li and Stubbins (2002) employed 2.0 mm wide, 7.9 mm long, and 0.8 mm thick miniature notch bar specimens and showed that three-point loading FCG tests produced similar crack growth behavior to that of standard specimens (CT) for three materials in yield strength (YS) ranging from 300 to 1200 MPa. Reliable crack growth data for a/w of 0.5 covering the full Paris law crack growth regime could be obtained.

The crack propagation data of Shin and Cai (2004) from small cylindrical specimens with lengths ranging from 22 to 200 mm and diameters of 8–15 mm under rotating bending, showed that specimens larger than 42 mm long yielded matching results with crack growth data of conventional tests. Further studies by Shin and Lin (2012) revealed that the miniature specimen crack growth rates (Figure 3.8) differed with the standard specimen

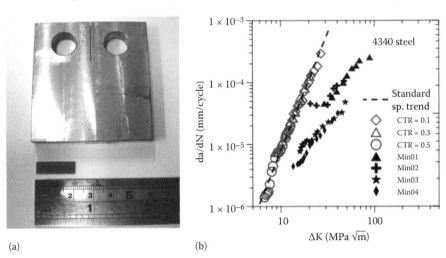

(a)   (b)

FIGURE 3.8
(a) Single-edge notch (SEN) miniature specimen of $20 \times 6 \times 0.5$ mm^3 compared with a standard CT specimen. (b) Comparison of FCG rates from the miniature specimen and standard CT specimen in terms of Delta K. (From Shin, C. S. and Lin, S. W., *International Journal of Fatigue* 43:105–110, 2012.)

results by a factor of 2–4 and, in most cases, the miniature specimen results were consistently below the standard specimen results. The discrepancy was attributed to the stress state effect, where a very thin specimen is in the plane stress state and exerts a smaller plastic constraint, resulting in a higher degree of crack closure as compared to a lower degree of crack closure, as well as a corresponding higher growth rate in the standard specimen that is near the plane strain state.

By employing a correction stress intensity range for closure effects, the miniature specimen data (single notch edge of 20 mm long by 6 mm wide by 0.5 mm thick) of steel and Al alloy could be made in line with the results of a standard 25 mm thick CT specimen. The growth rates versus the ΔK_{eff} relation from miniature specimen data can serve as an upper bound to get a conservative estimate of structural fatigue life.

3.5 Fracture Toughness from Small Punch Technique

The small punch (SP) has evolved as a promising technique not only for tensile property evaluation but also for derivation of toughness parameter. With decreasing test temperatures, load-displacement curves of SP test show a distinct fracture mode transition from ductile-to-brittle failure similar to that observed in the Charpy V-notch impact test. The SP experimental setup for evaluation of fracture properties are similar to those for defining the stress–strain behavior of materials discussed in the previous chapter.

3.5.1 Ductile–Brittle Transition from Small Punch

The approach for characterizing the ductile–brittle transition (Baik, Kameda, and Buck 1986; Kameda and Mao 1992) relies on the SP energy defined as the area under the load-displacement curves of the small punch test. The SP energy defined this way shows typical transition curve behavior when plotted against the test temperature (Figure 3.9). A small punch transition temperature is estimated by taking the temperature, defined as T_{sp}, at the midposition between the upper and lower shelf energies. The T_{sp} from SP tests are lower than the $(FATT)_{CVN}$ as the SP specimen are deformed under static loading and in a biaxial stress state, which is less favorable for brittle failure as compared with dynamic loading and triaxial stress state in CVN tests. Empirical correlations (Ha and Fleury 1998; Norris 1997; Matocha 2012) have been established between T_{sp} and FATT (both in kelvins) as

$$T_{sp} = \alpha \, (FATT)_{CVN} \tag{3.13}$$

where α is a material-specific constant in the range of 0.35–0.55.

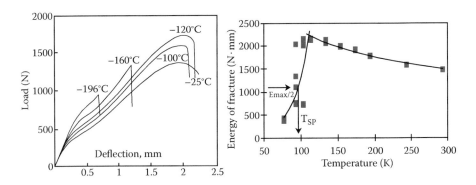

FIGURE 3.9
Typical SP curves at different test temperatures and the resulting energy versus temperature curve for T_{SP} determination.

Foulds and Viswanathan (1994) derived an empirical relationship of the form

$$\text{FATT (°C)} = A + B\, T_{sp}\ (\text{°C}) \tag{3.14}$$

where typical values of A are 2.3–2.8 and for B are in the range of 300–550 for various Cr–Mo steels. By performing multiple regression analyses over additional variables such as grain size, chemical composition, and phase percentage, Matsushita et al. (2000) showed that the correlation coefficient (R^2) was higher than that for the simple relation (as in Equation 3.13), thereby improving the reliability of transition temperature estimation of low alloy ferritic steels.

Matocha (2015) observed that the change in punch diameter from 2.5 to 2.0 mm does not significantly affect the temperature dependence of fracture energy in the transition region. However, the orientation of the SP test specimen affected the temperature dependence of fracture energy in the transition region and therefore the FATT-T_{sp} correlation.

3.5.2 Toughness—Lower and Upper Shelf

Energy-based and fracture stress formulations have been put forth to estimate the fracture toughness in the upper and lower shelf regimes (Hu and Fleury 1998). The methodology for fracture toughness estimation in the lower shelf is based on Ritchie, Knot, and Rice (RKR) criterion, which assumes that the cleavage fracture initiates ahead of a sharp crack when the applied load, σ_f, ahead of the crack tip over a characteristic distance, l_0, exceeds a critical value σ_f^*. The fracture toughness is given by the expression

$$K_{IC} = \beta^{-(n+1)/2}\, I_0^{1/2} \left(\frac{\sigma_f^{\left(\frac{n+1}{2}\right)}}{\sigma_y^{\left(\frac{n-1}{2}\right)}} \right) \tag{3.15}$$

where

β is a numerical constant obtained from the HRR (Hutchinson–Rice–Rosengren) small-scale yielding solution

n is the strain-hardening coefficient

σ_y is the yield stress

l_0 is the grain size

For low displacements in the SP sample, σ_{max} is computed as maximum elastic stress determined by Timoshenko analysis for a disk supported at the periphery and loaded in the center. The predictions of the fracture toughness in the brittle regime were found to be in good agreement with values obtained from standard CVN tests for various Cr-Mo steels.

In the ductile regime (upper shelf), Mao and Takahashi (1987) suggested that the failure of the SP specimen is controlled by the equivalent fracture strain under the biaxial stress state and derived an expression for plastic strain as

$$\varepsilon_q = \ln (t_0/t) \qquad (3.16)$$

where t_0 and t are original thickness and deformed thickness (near failure location), respectively. The biaxial fracture strain could be linearly related to elastic–plastic toughness as

$$J_{IC} = K\varepsilon_q - J \qquad (3.17)$$

where K and J are empirically determined constants.

Foulds et al. (1998) for the first time employed a video imaging system for identifying the onset of crack initiation and observed that crack initiates well in advance of the peak load and undergoes stable crack growth up to peak load in the ductile regime. Citing shortcomings in approaches based on equivalent fracture strain or peak load measurements, which assume crack initiation at peak loads, Foulds proposed an approach based on a continuum concept wherein the criterion for fracture is the strain energy density required for crack initiation in an uncracked SP specimen. This criterion determined using large strain FE analysis of SP loading is applied to the crack tip of a standard CT specimen. The strain energy density averaged over some distance x ahead of crack tip (W_{CT}) is computed. J_{IC} is then computed from the load level at which $W_{CT} = W_{SP}$. However, the key issue in this approach is the estimation of an appropriate averaging distance x, one of the procedures based on the blunting crack extension (Δa). Using this method, fracture toughness of a wide variety of power plant steels could be predicted within an accuracy of ±25%.

3.5.3 Small Punch Using Notched Specimen

Most early research on small punch (SP) employed uncracked specimens for deriving the fracture properties using the area under the load-displacement curve or the displacement at failure. Analogous to the conventional fracture toughness evaluation procedures that use notched Charpy and precracked single-edge notched bending (SENB) or CT specimens, the use of precracked SP specimens was attempted to assure a more solid basis for the estimation of transition temperature and fracture toughness parameters.

One of the earliest studies was performed by Ju, Jang, and Kwon (2003) using precracked SP specimens in order to evaluate the fracture properties of different steels. A sharp notch was introduced in the center of the SP specimen as shown in Figure 3.10a and the stress intensity factor, K, of the cracked thin plate under bending force was derived as

$$K_C = \frac{3}{2}\frac{P(1+v)}{\pi h^2}\left(\ln\left(\frac{c}{b}\right) + \frac{b^2}{4c^2}\right)\sqrt{a} \tag{3.18}$$

where b and c are the contact radius and the lower die radius, respectively, P is the crack initiation point, and a is the half of the initial crack opening. Determining the crack initiation point using acoustic emission signals and b as a function of applied load, the fracture toughness of SA 508 steel in the lower shelf compared well with ASTM E1921 master curve results. However,

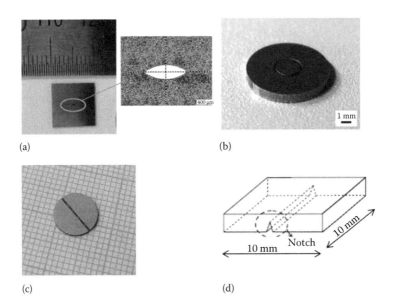

(a)   (b)

(c)   (d)

FIGURE 3.10
Various notched SP specimen geometry. (a) A through thickness-type notch (1.0 mm length and 0.5 mm wide) employed by Ju, Jang, and Kwon (2003); (b) sharp circular notch studied by Turba et al. (2011); (c) disc specimen with U notch in the axis of disc plane (Matocha 2015); and (d) notched SP specimen studied by Cardenas et al. (2012).

the room temperature toughness values were much lower than the master curve results, possibly due to noninclusion of plastic deformation and crack extension in the analysis of Equation 3.18.

A circular notch of 2.5 mm diameter in a disk-shaped specimen of 8.0 mm diameter (Figure 3.10b) was employed by Turba et al. (2011). The depth of the notch was equal to 0.5 mm in a specimen with a thickness of 1 mm. Though this geometry markedly decreased both the displacement at fracture and the SP fracture energies, the shift in T_{SP} toward the $(DBTT)_{CVN}$ was not seen, leading to a conclusion that the observed shift between T_{SP} and $(DBTT)_{CVN}$ seems to be related to a combination of the strain rate and size effects rather than the presence of a sharp notch. However, disk specimen with a U-shaped notch in the axis plane of the disk (Figure 3.10c) shifts transition temperature T_{SP} significantly to higher temperatures and the temperature dependence of fracture energy was also less steep (Matocha 2015).

Different notch configurations in a squared SP specimen, such as a longitudinal notch (L), longitudinal + transverse notch (L+T), and circular notch, were studied by Cardenas et al. (2012); based on the trends of stress triaxiality with relative depth of the notch, the L configuration (Figure 3.10d) with a notch depth-to-thickness ratio a/t = 0.4 was chosen for detailed study. The load at the onset of crack initiation was experimentally determined by interrupted tests with the aid of scanning electron microscope observations. With this input, the test was modeled using finite element analysis with ductile damage models and the J-integral was evaluated as a contour integral in ABAQUS. The J_{SPT} ($\Delta a = 0$) for actual crack initiation in a notched SPT sample obtained by this method was higher than that of the standard J, possibly due to the loss of constraint resulting from the low thickness of the SPT sample as well as to the use of a blunt notch.

The notch configuration employed by Lacalle et al. (2012) is lateral through thickness notch (Figure 3.10e) with initial crack length of about 4.5–5.0 mm from the sample edge. The value of notch opening displacement corresponding to the crack initiation event is determined by numerical simulation using the punch displacement (δ) from experiments and transformed into K or J using the classical principles. The J_c from SP tests for different steels compared well with that of conventional tests with confidence margins of ±15%.

3.6 Fracture Toughness from Indentation Techniques

The load-indentation depth data obtained in a ball indentation test can also be used for evaluation of fracture toughness of the material. Various methods developed for fracture toughness estimation from indentation are presented here.

3.6.1 Indentation Energy to Fracture

The concept of indentation energy to fracture (IEF) allows the nondestructive determination of fracture energy from automated ball indentation (ABI)-measured true stress–strain curves based on a premise that concentrated stress fields near and ahead of the indenter-specimen contact are similar to that ahead of a crack except that the indentation stress fields are compressive (Haggag et al. 1998). IEF is defined as

$$\text{IEF} = \int_0^{h_f} P_m(h)\,dh \tag{3.19}$$

where

$P_m = \dfrac{4P}{d^2}$ is the mean contact pressure

P is the indentation load
h is the indentation depth
h_f is the indentation depth up to cleavage fracture stress
d is the chordal diameter of the indentation

The slope (S) of the load-depth curve is used to calculate IEF as

$$\text{IEF} = \frac{S}{\pi}\ln\left(\frac{D}{D-h_f}\right) \tag{3.20}$$

In absence of critical fracture stress for a material, a reference stress was employed by Haggag and colleagues to compute h_f. The IEF computed was seen to depict the ductile–brittle transition temperature for carbon steels.

Byun, Kim, and Hong (1998) furthered this IEF concept and proposed a theoretical model to estimate the fracture toughness of ferritic steels in the transition region. Dividing the fracture energy per unit area into two terms: a temperature-independent term, W_0 (= lower shelf energy), and a temperature-dependent term, (W_T), the key assumptions in this model were that (1) the indentation energy per unit contact area is identical to the temperature-dependent term $(W_{IEF} = W_T)$, and (2) fracture occurs when the maximum contact pressure reaches the fracture stress of the material. The fracture stress was estimated from ABI test data by coupling the IEF theory with fracture mechanics models and using the lower shelf fracture toughness data. By performing ABI tests on RPV steels at temperatures from −150°C to ambient, the temperature dependence of the estimated K_{JC} was found to be same as that of the ASTM K_{JC} master curve.

3.6.2 Continuum Damage Mechanics Approach

For estimating fracture toughness of ductile materials, Lee et al. (2006) proposed a critical indentation energy model exploiting the basic concepts of continuum damage mechanics (CDM). Indentation energy per unit contact area is related to w_f, the energy per unit area required to create a crack surface that is expressed as

$$2w_f = \lim_{h \to h^*} \int_0^h \frac{4P}{\pi d^2}\, dh \qquad (3.21)$$

where
  P is the applied indentation load
  h is the indentation depth
  d is the impression diameter
  h^* is the critical indentation depth corresponding to the characteristic fracture initiation point.

This characteristic point was determined using the concept of damage variable and its relation to the elastic modulus of the damaged material (E_D). The localized shear stresses due to compressive indentation induce void nucleation, and increase in void volume with penetration depth causes E_D to decrease with depth. Expressing E_D as a function of indentation parameters and using a critical void volume fraction of 0.25 for crack growth, the depth corresponding to the critical E_D^* was estimated and used for computing w_f and K_{JC}. The estimated fracture toughness values obtained from the indentation technique showed good agreement within 10% error with those from conventional crack tip opening displacement (CTOD) tests.

3.7 Concluding Remarks

The main factor to be considered in the evaluation of toughness is the reduced constraints in miniaturized specimens as compared to their conventional counterparts. The use of subsize Charpy impact tests relies heavily on the correlation with full size specimen tests, while for fracture toughness methods, the underlying source of size effects needs to be addressed, for example, through a master curve approach, which enables the use of smaller size specimens. A better understanding of the constraint effects on fracture toughness is imperative to bring the small specimen methodologies under the ambit of standards and codes for its acceptance. Integration of small specimen tests with numerical analysis and fracture models together with microstructural analysis is the way forward.

References

Baik, J. M., Kameda, J., and Buck, O. 1986. Development of small punch tests for ductile brittle transition temperature measurement of temper embrittled NiCr steels. In *The use of small scale specimens for testing irradiated specimens*, STP 888, eds. W. R. Corwin and G. E. Lucas, 92–111. Philadelphia: ASTM.

Bohme, W., and Schmitt W. 1998. Comparison of results of instrumented Charpy and mini-Charpy tests with different RPV steels. In *Small specimen test techniques*, ASTM STP 1329, eds. W. R. Corwin, S. T. Rosinski, and E. van Walle, 32–47. Philadelphia: American Society for Testing and Materials.

Byun, T. S., Kim, J. W., and Hong, J. H. 1998. A theoretical model for determination of fracture toughness of reactor pressure vessel steels in the transition region from automated ball indentation test. *Journal of Nuclear Materials* 252: 187–194.

Cardenas, E. C., Belzunce, F. J., Rodriguez, C., Penuelas, I., and Betegon, C. 2012. Application of the small punch test to determine the fracture toughness of metallic materials. *Fatigue and Fracture of Engineering Materials* 35:441–450.

Corwin, W. R. and Hougland, A. M. 1986. Effect of specimen size and material condition on the Charpy impact properties of 9Cr-1Mo-V-Nb steel. In *The use of small-scale specimens for testing irradiated material*, ASTM STP 888, eds. W. R. Corwin, S. T. Rosinski, and E. van Walle, 325–338. Philadelphia: American Society for Testing and Materials.

Corwin, W. R., Klueh, R. L., and Vitek, J. M. 1984. Effect of specimen size and nickel content on the impact properties of 12 Cr-1 MoVW ferritic steel. *Journal of Nuclear Materials* 122:343–348.

Ermi, A. M. and James, L. A. 1986. Miniature center-cracked-tension specimen for fatigue crack growth testing. In *The use of small-scale specimens for testing irradiated material*, ASTM STP 888, eds. W. R. Corwin and G. E. Lucas, 261–275. Philadelphia: American Society for Testing and Materials.

Foulds, J. R. and Viswanathan, R. 1994. Small punch testing for determining the material toughness of low alloy steel components in service. *Journal of Engineering Materials Technology* 116:457–464.

Foulds, J. R., Wu, M., Srivastav, S., and Jewett, C. W. 1998. Fracture and tensile properties of ASTM cross-comparison exercise on A533B steel by small punch testing. In *Small specimen test techniques*, STP 1329, eds. W. R. Corwin, S. T. Rosinski, and E. van Walle, 557–574. Philadelphia: ASTM.

Ha, J. S. and Fleury, E. 1998. Small punch tests to estimate the mechanical properties of steels for steam power plant. II. Fracture toughness. *International Journal of Pressure Vessels and Piping* 75:707–713.

Haggag, F. M., Byun, T. S., Hong, J. H., Miraglia, P. Q., and Murty, K. L. 1998. Indentation-energy-to-fracture (*IEF*) parameter for characterization of DBTT in carbon steels using nondestructive automated ball indentation (ABI) technique. *Scripta Materialia* 38:645–651.

Hirose, T., Sakasegawa, H., Kohyama, A., Katoh, Y., and Tanigawa, H. 2001. Effect of specimen size on fatigue properties of reduced activation ferritic/martensitic steel. In *ASTM STP 1405*, 535. West Conshohocken, PA: American Society for Testing and Materials.

Huang, F. 1986. Use of subsized specimens for evaluating the fracture toughness of irradiated materials. In *The use of small-scale specimens for testing irradiated material*, ASTM STP 888, eds. W. R. Corwin and G. E. Lucas, 290–304. Philadelphia: American Society for Testing and Materials.

Ju, J. B., Jang, J. I., and Kwon, D. 2003. Evaluation of fracture toughness by small-punch testing techniques using sharp notched specimens. *International Journal of Pressure Vessels and Piping* 80:221–228.

Kameda, J. and Mao, X. 1992. Small-punch and TEM-disc testing techniques and their application to characterization of radiation damage. *Journal of Materials Science* 27:983–989.

Kayano, H., Kurishita, H., Kimura, A., Narui, M., Yamazaki, M., and Suzuki, Y. 1991. Charpy impact testing using miniature specimens and its application to the study of irradiation behavior of low-activation ferritic steels. *Journal of Nuclear Materials* 179–181:425–428.

Korolev, Y. N., Kryukov, A. M., Nikolaev, Y. A., Platonov, P. A., Shtrombakh, Y. I., Langer, R., Leitz, C., and Rieg, C.-Y. 1998. The actual properties of WWER-440 reactor pressure vessel materials obtained by impact tests of subsize specimens fabricated out of samples taken from the RPV. In *Small specimen test techniques*, ASTM STP 1329, eds. W. R. Corwin, S. T. Rosinski, and E. van Walle, 145–159. Philadelphia: American Society for Testing and Materials.

Kumar, A. S., Garner, F. A., and Hamilton, M. L. 1990. Effect of specimen size on the upper shelf energy of ferritic steels. In *Effects of radiation on materials: 14th International Symposium (vol. II)*, ASTM STP 1046, eds. N. H. Packan, R. E. Stoller, and A. S. Kumar, 487–495. Philadelphia: American Society for Testing and Materials.

Kumar, A. S., Louden, B. S., Garner, F. A., and Hamilton, M. L. 1993. Recent improvements in size effects correlations for DBTT and upper shelf energy of ferritic steels. In *Small specimen test techniques applied to nuclear reactor vessel thermal annealing and plant life extension*, ASTM STP 1204, eds. W. R. Corwin, F. M. Haggag, and W. L. Server, 47–61. Philadelphia: American Society for Testing and Materials.

Kurishita, H., Yamamoto, T., Narui, M., Suwarno, H., Yoshitake, T., Yano, Y., Yamazaki, M., and Matsui, H. 2004. Specimen size effects on ductile–brittle transition temperature in Charpy impact testing. *Journal of Nuclear Materials* 329–333:1107–1112.

Lacalle, R., Alvarez, J. A., Arroyo, B., and Solana, F. G. 2012. Methodology for fracture toughness estimation based on the use of small punch notched specimens and CTOD concept. In *Proceeding of the II International Conference SSTT*, eds. K. Matocha, R. Hurst, and W. Sun, 171–177. Ostrava, Czech Republic.

Lee, J. S., Jang, J. I., Lee, B. W., Choi, Y., Lee, S. G., and Kwon, D. 2006. An instrumented indentation technique for estimating fracture toughness of ductile materials: A critical indentation energy model based on continuum damage mechanics. *Acta Materialia* 54:1101–1109.

Li, M., and Stubbins, J. F. 2002. Subsize specimens for fatigue crack growth rate testing of metallic materials. In *Small specimen test techniques*, STP 1418, eds. M. A. Sokolov et al., 321–328. West Conshohocken, PA: ASTM International.

Liu, K. C., and Grossbeck, M. L. 1986. Use of subsize fatigue specimens for reactor irradiation testing. In *The use of small-scale specimens for testing irradiated material*, ASTM STP 888, eds. W. R. Corwin and G. E. Lucas, 276–289. Philadelphia: American Society for Testing and Materials.

Louden, B. S., Kumar, A. S., Garner, F. A., Hamilton, M. L., and Hu, W. L. 1988. The influence of specimen size on Charpy impact testing of unirradiated HT-9. *Journal of Nuclear Materials* 155:662–667.

Lucas, G. E., Odette, G. R., Matsui, H., Moslang, A., Spatig, P., Rensman, J., and Yamamoto, T. 2007. The role of small specimen test technology in fusion materials development. *Journal of Nuclear Materials* 367–370, 1549–1556.

Lucas, G. E., Odette, G. R., Sheckherd, J. W., McConnell, P., and Perrin, J. 1986. Subsized bend and Charpy V-notch specimens for irradiated testing. In *The Use of small-scale specimens for testing irradiated material,* ASTM STP 888, eds. W. R. Corwin and G. E. Lucas, 304–324. Philadelphia: American Society for Testing and Materials.

Lucas, G. E., Odette, G. R., Sokolov, M., Spatig, P., Yamamoto, T., and Jung, P. 2002. Recent progress in small specimen test technology. *Journal of Nuclear Materials* 307–311:1600–1608.

Lucon, E. 1998. Subsize impact testing: CISE experience and the activity of the ESIS TC 5 Sub-committee. In *Small specimen test techniques,* ASTM STP 1329, eds. W. R. Corwin, S. T. Rosiinski, and E. van Walle, 15–31. West Conshohocken, PA: ASTM International.

Mao, X. 1991. Fracture toughness: JIC prediction from super-small specimens (0.2 CT, 0.5 mm thick) of a martensitic stainless steel HT-9. *Journal of Engineering Materials Technology* 113 (1):135–140.

Mao, X. and Takahashi, H. 1987. Development of a further miniaturized specimen of 3 mm diameter for TEM disk (∅ 3 mm) small punch tests. *Journal of Nuclear Materials* 150:42–52.

Manahan, M. P., Sr. 1999. In situ heating and cooling of Charpy test specimens. In *Pendulum impact testing: A century of progress*, ASTM STP 1380, eds. T. A. Siewert and M. P. Manahan, Sr., 286–297. West Conshohocken, PA: American Society for Testing and Materials.

Matocha, K. 2012. Determination of actual tensile and fracture characteristics of critical components of industrial plants under long term operation by SPT. In *Proceedings of the ASME 2012 Pressure & Piping Division Conference PVP 2012.* Toronto, Ontario, Canada.

Matocha, K. 2015. Small punch testing for tensile and fracture behaviour—Experiences and way forward. In *Small specimen test techniques*, STP 1576, eds. S. Mikhail and L. Enrico, 145–159, West Conshohocken, PA: ASTM International, doi:10.1520 /STP IS7620140005.

Matsushita, T., Saucedo, M. L., Joo, Y. H., and Shoji T. 2000. Development of a multiple linear regression model to estimate the ductile–brittle transition temperature of ferritic low-alloy steels based on the relationship between small punch and Charpy V-notch tests. *Journal of Testing and Evaluation* 28(5).

Moitra, A., Krishnan, S. A., Sasikala, G., Bhaduri, A. K., and Jayakumar, T. 2015. Effect of specimen size in determining ductile to brittle transition temperature. *Materials Science and Technology* 31:1781–1787.

Moslang, A. 2000. Development of creep fatigue specimen and related test technology. IFMIF User Meeting. Tokyo, Japan.

Norris, S. D. 1997. A comparison of the disk bend test and the charpy impact test for fracture property evaluation of power station steels. *International Journal of Pressure and Vessel Piping* 74:135–144.

Odette, G. R., Edsinger, K., Lucas, G. E., and Donahue, E. 1998. Development of fracture assessment methods for fusion reactor materials with small specimens. In *Small specimen test techniques*, STP 1329, eds. W. R. Corwin, S. T. Rosinski, and E. van Walle, 298–327. Philadelphia: ASTM.

Ono, H., Kasada, R., and Kimura, A. 2004. Specimen size effects on fracture toughness of JLF-1 reduced-activation ferritic steel. *Journal of Nuclear Materials* 329–333:1117–1121.

Pahlavanyali, S., Rayment, A., Roebuck, B., Drew, G., and Rae, C. M. F. 2008. Thermo-mechanical fatigue testing of superalloys using miniature specimens. *International Journal of Fatigue* 30 (2):397–403.

Schindler, H. J. and Veidt, M. 1998. Fracture toughness evaluation from instrumented subsize Charpy type tests. In *Small specimen test techniques*, ASTM STP 1329, eds. W. R. Corwin, S. T. Rosinski, and E. van Walle, 48–62. Philadelphia: American Society for Testing and Materials.

Schubert, E., Kumar, A. S., Rosinski, S. T., and Hamilton, M. L. 1995. Effect of specimen size on the impact properties of neutron irradiated A533B steel. *Journal of Nuclear Materials* 225:231–237.

Shin, C. S. and Chen, P. C. 2004. Fatigue crack propagation testing using subsized rotating bending specimens. *Nuclear Engineering and Design* 231:13–26.

Shin, C. S. and Lin, S. W. 2012. Evaluating fatigue crack propagation properties using miniature specimens. *International Journal of Fatigue* 43:105–110.

Sokolov, M. A. and Nanstad, R. K. 1995. On impact testing of subsize Charpy V-notch type specimens. In *Effects of radiation on materials (vol. 17)*, ASTM STP1270. D. S. Gelles, R. K. Nanstad, A. S. Kumar, and E. A. Little, 384–414. Philadelphia: American Society for Testing and Materials.

Tomimatsu, M., Kawaguchi, S., and Iida, M. 1998. Reconstitution of fracture toughness specimens for surveillance test. In *Small specimen test techniques*, ASTM STP 1329, eds. W. R. Corwin, S. T. Rosinski, and E. van Walle, 470–483. Philadelphia: ASTM.

Turba, K., Gülçimen, B., Li, Y. Z., Blagoeva, D., Hähner, P., and Hurst, R. C. 2011. Introduction of a new notched specimen geometry to determine fracture properties by small punch testing. *Engineering Fracture Mechanics* 78:2826–2833.

van Walle, E., Scibetta, M., Valo, M., Viehrig, H. W., Richter, H., Atkins, T., Wootton, M., Keim, E., Debarberis, L., and Horsten, M. 2002. RESQUE: Reconstitution techniques qualification and evaluation to study ageing phenomena of nuclear pressure vessel materials. In *Small specimen test techniques*, STP 1418, eds. M. A. Sokolov et al., 409–425. West Conshohocken, PA: ASTM International.

Viehrig, H. W. and Boehmert J. 1998. Specimen reconstitution technique and verification testing for Charpy size SENB specimens. In *Small specimen test techniques*, STP 1329, eds. W. R. Corwin, S. T. Rosinski, and E. van Walle, 420–435. Philadelphia: ASTM.

Wallin, K. 1999. The master curve method: A new concept for brittle fracture. *Journal of Materials and Product Technology* 14:342–354.

Wallin, K., Planman, T., Valo, M., and Rintamaa, R. 2001. Applicability of miniature size bend specimens to determine the master curve reference temperature T_0. *Engineering Fracture Mechanics* 68:1265–1296.

4

Critical Issues in Small Specimen Testing

4.1 Introduction

The characterization of mechanical properties with small size specimens involves the following issues: (1) size effects and scaling to bulk material behavior; (2) specimen design and preparation; (3) tooling, force application, and strain measurements; and (4) optimized procedures leading to standards.

4.2 Specimen Size Effects and Their Influence on Mechanical Behavior

It is well known that the strength of metallic alloys is governed by microstructural constituents such as dislocations, grains, second-phase particles, etc. In case of conventional specimen designs with sizes of a few tens of millimeters, sufficient numbers of such microstructural features participate in the deformation. The measured properties in such cases are generally independent of specimen size and are representative of the bulk material behavior.

Mechanical properties change drastically when the specimen dimensions are small and the term "size effect" is used generically to describe this. The size effect manifests as the specimen dimensions approach the length scales of the defects controlling the deformation where the laws of continuum mechanics are no longer valid. This is true for materials in miniaturized devices used in the form of thin films, whose dimensions become comparable to the microstructural length scale. In such cases, the aim is to evaluate the mechanical response of materials at length scales comparable to those used in actual application. The small size specimens are close to reality when coatings, thin films, microelectromechanical structure, or materials for such applications are tested and evaluated, as illustrated in Figure 4.1.

Size effects are categorized as "intrinsic" or "extrinsic" (Zhu, Bushby, and Dunstan 2008). Intrinsic size effects arise due to microstructural constraints,

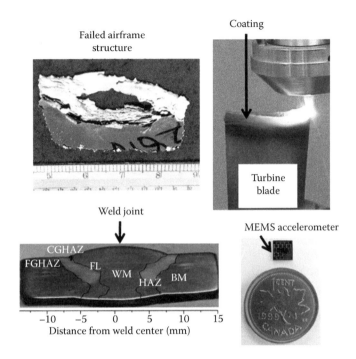

FIGURE 4.1
Examples of situations where testing at small scales is inevitable.

such as grain size or second-phase particles/precipitates. The strength of polycrystalline metals is inversely proportional to grain size according to the well known Hall–Petch relation, which breaks down for nanometer size grains. Extrinsic size effects are caused by dimensional constraints due to small sample size, where dislocation motion and other physical mechanisms are affected by the presence of a surface or interface, or due to small strained volume. Therefore, mechanical properties of materials evaluated in small volumes may differ significantly from those evaluated in bulk form due to both geometric and microstructural constraints (Nagamani and Alam 2013).

For mechanical tests with miniaturized specimens in the size range of 0.5–5.0 mm to represent bulk material behavior, the various forms of size effect to be considered include

1. The requirement to have sufficient representative volume elements (grains, colonies) within the deforming volume for stress–strain measurements
2. The necessity to ensure the effects of reduced constraint around a notch for toughness evaluation
3. Relevance of deformation and failure models and the likelihood of probabilistic events from smaller numbers of defects on deformation and failure

One of the approaches for bridging deformation across different length scales is based on the concept of representative volume element (RVE). RVE is the volume of the material microconstituents and microstructures that can be correlated to the properties of macroscopic materials. RVE is usually regarded as a volume, V, that includes a sample of all microstructural heterogeneities and is sufficiently large to be statistically representative of the material (Figure 4.2). The minimum RVE size is not only material dependent, but also material property dependent.

Ren and Zheng (2002, 2004) performed numerical experiments using the finite element method (FEM) to determine the minimum RVE sizes of more than 500 cubic polycrystals and established that minimum RVE sizes for effective elastic moduli were 16 times the grain size. Though numerical studies to characterize RVEs based on microstructure and/or elastic deformation have been reported, experimental measurements to determine what constitutes an RVE for plastic deformation are very scarce. The length scale of RVE of a plastically deforming isotropic cubic polycrystal has been determined as approximately two times the grain size (Ranganathan and Ostoja 2008). For hexagonal crystal systems such as titanium subjected to cyclic tensile loading, RVE of 3.3 times the average grain diameter has been reported based on multiscale strain optical image measurements (Efstathiou, Sehitoglu, and Lambros 2009). In the case of a polycrystalline aggregate of 316L steel with a crack under tensile load, Simonovski and Cizelj (2007) numerically estimated RVE based on crystal plasticity and material models. By studying the crack tip opening displacements (CTODs) as a function of different sizes of polycrystalline aggregates with number of grains ranging from 145 to 5027, their study concluded that RVE size is practically 4 × 3 mm for a cracked SS316L.

For conventional bulk specimens, a large number of grains are involved in the deformation process and the obtained properties are a combined result of different grains and random distributions of these grain orientations. On

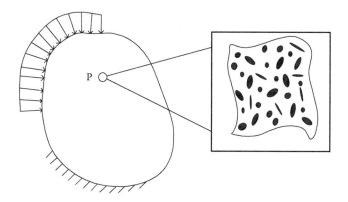

FIGURE 4.2
Schematic illustrating RVE.

the contrary, for small-scale specimens (e.g., low thickness, larger grain size), the role of individual grain become increasingly significant to the overall deformation behavior. Size effects have been studied by examining the interplay of grain size (d_g), specimen shape, and thickness (t) on the mechanical properties. When the ratio t/d_g is smaller than a certain value, the mechanical behavior of small size specimens is found to be substantially different from that of standard specimens. The determination of the critical t/d_g is important from the viewpoint of engineering application of the small specimen tests for predicting the bulk material behavior.

4.2.1 Size Effect in Subsize Tensile and Punch Tests

In the case of uniaxial straining, most of the studies have shown that a thickness-to-grain size ratio (t/d_g) of greater than 6–10 gives representative results of the bulk. The critical thickness-to-grain size ratio, however, depends on the particular material under investigation; for example, critical t/d_g for austenitic stainless steel 316 is 6 (Igata et al. 1986). Studies by Yang and Lu (2013) on tensile properties of pure copper show that the critical value of t/d_g depends on grain size as well; for example, the critical t/d_g is about 6 for coarse-grained Cu ($d_g \sim 82$ µm), while that for fine-grained Cu of grain size 13 µm is around 18. When t/d_g is below the critical value, the strength and elongation decrease, implying that "smaller is weaker." Thickness effects in this study have been explained based on differences between the dislocation activities in the interior grains and surface grains. The grain size effects were further studied by Janssen, de Keijser, and Geers (2006) in the limit $t = d_g$ for polycrystalline 99.999 at.% aluminum sheet with thickness (t) of 100–340 µm and grain size between 75 and 480 µm. The study indicated a reverse trend (i.e., strength increase when specimen thickness contained crystals with two or fewer free surfaces).

Size effect studies in shear punch and small punch (SP) test configurations have been rather limited. Toloczko et al. (2002) analyzed shear punch data for various t/d_g between 3.1 and 18 for AISI 316 steels by varying specimen thickness from 0.1 to 0.3 mm and for grain sizes of 17 and 31 µm. They observed that there was no significant effect of t/d_g on the shear yield and maximum strengths from shear punch tests of AISI 316 steels for t/d_g in a range of 3.1–18, in contrast to dependence of tensile yield strength (YS) on $t/d_g < 6$. According to Toloczko and colleagues, the reduced sensitivity of shear punch test data to t/d_g was likely due to a combination of mechanical constraint in the shear punch loading configuration and the numbers of grains in the clearance zone participating in the deformation.

Another notable parameter leading to size effect in shear punch test is the die-punch clearance (the annular region between the flat punch and lower die) where the shear dominant deformation takes place. Using finite element analysis, Guduru et al. (2007) examined the role of specimen thickness and clearance zone width (c) for a range of thickness from 0.25 to 0.8 mm and (c) ranging from 5 to 20 μm for three different materials (Nb, brass, and steel). For a fixed die-punch clearance, decreasing specimen thickness increases the relative contribution of bending stress and hence decreases the stress for yielding. Similarly, for a fixed specimen thickness, increasing the die-punch clearance enhances the bending stress contribution, decreasing the stress required for yielding. However, for the range of specimen thickness and clearance zone width examined by Guduru et al., the variations in shear yield strength were very small, within the scatter observed in their experimental values.

Song et al. (2014) studied the size effects in an SP test of AISI 316 steels for grain size in the range of 10–50 μm and thickness in the range of 0.1–0.5 mm. Their study revealed that critical value of t/d_g is about 19–25 above which the strength dependencies followed the Hall–Petch relationship and that the correlation of SP yield and maximum loads with conventional YS and ultimate tensile strength (UTS), respectively, were constant. There was no obvious size effect in the samples with smaller grain size (10, 20, and 30 μm), while this effect existed in the samples with larger grain size (40 and 50 μm) and thin specimen thickness (0.1 mm). The size effects in SP test causing a YS increase with decreasing thickness/increasing grain size (i.e., t/d_g) were explained in terms of the restrained grain interaction with increasing grain size leading to the decrease in dislocation sources. The deformation behavior of very thin samples with relatively coarse grains becomes anisotropic and is affected by the neighboring grain orientations leading to the observed size effects. It may be noted that critical (t/d_g) for a shear punch test is much less than small punch loading due to differences in loading configurations and stress state during the specimen deformation in the two test techniques.

4.2.2 Specimen Size Requirements for Spherical Indentation

The rules for specimen dimensions for spherical indentation tests are the same as that set forth for hardness testing as per ASTM E 10 and E 384. The thickness (t) of the sample is generally maintained at least 10 times the penetration depth to ensure that the plastic zone is contained well within the volume of the material indented and is not affected by the back surface. When this limit is breached, the data is no longer representative of the material. According to Lucas, Sheckherd, and Odette (1986), for small-diameter indenters, the plastic zone initially forms close to the specimen surface and becomes fully developed before being affected by the back surface. This limits the indentation diameters to less than 0.25–0.5t. For large-diameter indenters, the plastic zone may initiate far from the top surface, but may be affected by back surface before it develops fully. Based on these considerations, the minimum thickness proposed by Lucas was expressed as

$$t_{min} = \min (2 - 4\, d_{max},\ 10\, h_{max})$$

where d_{max} is the maximum indentation diameter and h_{max} is the maximum penetration depth.

4.2.2.1 Indentation Size Effect

Tabor (1954) showed that, for a given metal, the indentation stress–strain curve is independent of ball size for large-radius indenters. Several studies have shown that for smaller radius spherical indenters (e.g., R < 100 μm) the indentation stress–strain curve shifts to increasingly higher pressures. While for pointed indenters of fixed geometry (i.e., Berkovich, Vickers, and cones), the hardness is seen to increase as the depth of penetration decreases, spherical indenters show a dependence of hardness on the radius of the indenter rather than on the penetration depth (Swadener, George, and Pharr 2001).

The size effect has been simulated with small radius indenters by Spary, Bushby, and Jennett (2006) through an increase in the initial yield stress, but keeping the work-hardening rate the same as that of bulk indentation experiments. Most studies have shown that indentation size effect is a geometrical effect linked to the stress distribution under the indenter for the initiation of yielding over a finite volume. Strength increase is believed to be associated with the difficulty in initiating plasticity within the steep elastic strain gradients, resulting in higher densities of geometrically necessary dislocations beneath a small spherical indenter.

4.2.3 Size Effect in Fracture Toughness Testing

It is well known that fracture toughness depends on the test specimen size and geometry, and there is an increase in toughness with decreasing specimen thickness associated with the transition from conditions of plane strain to plane stress. This form of size effect is related to the stress and stress-state distributions in the local volume of material near the tip of a blunting crack. This effect, known as a constraint loss effect, is primarily due to the size and geometry factors that reduce the amplitude of the crack tip stress fields below small-scale yielding (SSY) levels. The constraint is influenced by the plastic zone size around the notch relative to the specimen dimensions, which is expressed as a function of the local yield stress, the remaining ligament, and the thickness.

The possibility to derive toughness from small specimens is more restrictive as compared to the flow properties because size effect is linked to geometrical parameters involving plastic zone size, which are much larger than the microstructural elements. The specimens for fracture toughness evaluation have a size criterion dictated by the constraint to produce a size-independent toughness value. However, the size considerations are greatly reduced in materials of high yield strength and low ductility (brittle), where reduced plastic zone size brings down the minimum thickness required for valid toughness.

For cleavage types of fractures, in addition to the constraint effect, there exists an inherent size effect due to variations in the highly stressed volume of material near the crack tip that affects the measured toughness. With decreasing specimen thickness, the stressed volume decreases, leading to a reduced statistical probability that a sufficiently high stress will encounter a sufficiently weak trigger particle (large grain boundary carbides, or clusters of carbides, and other brittle inclusions), producing a propagating microcrack resulting in macroscopic cleavage. The probability of triggering a large, weak microcrack nucleation site not only varies with the stressed volume, but also from specimen to specimen, resulting in scatter in measured cleavage fracture toughness. The Weibull statistical model of fracture, which proposes strength as a function of volume of defects, is generally incorporated for size adjustments in the analysis of small-scale toughness data (Rathbun et al. 2006).

4.3 Issues Related to Specimen Orientation and Stress State

In mechanical tests with scaled-down subsize specimens, the loading configuration is similar to the respective conventional equivalent. The specimen orientation with respect to certain material characteristics (such as grain shape or texture) is the same in both the tests, making the interpretation and scaling to bulk relatively straightforward.

In techniques such as small punch, shear punch, and ball-indentation (BI) tests, the loading configuration is biaxial, shear, and triaxial compression, respectively—different from the standard tensile (uniaxial loading) and fracture toughness (bend bar, compact tension, center-cracked panel) tests with which they are correlated. Specimen orientation is of no consideration as long as the material behavior is isotropic. However, when the material response is anisotropic, such as in the case with high grain aspect ratios (e.g., rolled product) textured microstructure (including directionally solidified material), due consideration of specimen orientation in the small specimen test is very crucial to obtain representative results comparable to conventional tests.

The code of practice of the European Committee for Standardization (CEN) WS (2007) specifies certain guidelines for specimen orientation in SP tests employed for assessment of tensile properties. In the case of a rolled product, it is recommended that the SP test specimen be oriented so that the specimen disk plane is coincident with the isotropic plane (i.e., transverse plane of rolled product). In situations where there is no plane of isotropy, specimen orientation should be chosen so that the desired direction of uniaxial testing is in the SP specimen disk plane. Similarly, for standard bend-type fracture specimen equivalence, the SP specimen disk plane is normal to the direction of crack advance in the fracture specimen.

The use of spherical indentation to evaluate stress–strain properties is based on the assumption that properties in tension and compression are identical and that material behavior is isotropic. The load-depth traces does not reflect the anisotropic nature of the material, while an asymmetry in the residual imprint is indicative of the specimen anisotropy. To characterize anisotropic material using spherical indentation, attempts have been made to modify the von Mises yield criterion to account for the anisotropy of the deformation process (Oviasuyi 2012). The indented material deforms plastically in the axial direction of indentation, but also deforms laterally in a plane that includes the radial and the transverse directions of the material. Applying Hill's (1971) anisotropy criterion and determining the constants F, G, and H from compression tests, the fractional constraint to the plastic deformation occurring from indentation stress in each of the orthogonal directions of the sample is determined. The total equivalent yield stress can be calculated by combining these fractional contributions to determine the constraint factor and the average stress. Using this methodology, Oviasuyi (2012) showed that the flow curves from BI tests of Zr–2.5%Nb reactor pressure tube material were found to be very similar to those obtained from conventional uniaxial compression and tension tests.

4.4 Specimen Preparation Methods

Specimen preparation methods for small specimen testing depend on the type of test to be carried out and the material examined. For specimen sizes in the range of 0.5–5.0 mm for subsize tensile and punch/indentation tests (disk specimens), the most common method of specimen preparation from metallic alloys involves removing material from the component by cutting and machining to the required size. All cutting processes produce a damaged layer; for example, a cut-off wheel produces about a 0.5 mm layer of damage near the surface, which needs to be removed by polishing or careful diamond grinding.

A machining process that does not introduce any extraneous stresses or alterations in the microstructure is desirable. The most popular cutting method adopted for small specimen preparation is the computer controlled electric discharge machining (EDM). EDM uses an electrostatic discharge between the electrode and workpiece to remove material. A dielectric fluid is used to flush away debris and act as coolant in the erosive technique. The cutting process by EDM has advantages of high precision and low cutting loss and it is amenable to programming and automation. The recast layer and heat-affected zone, typically about 0.030–0.050 mm thick, can be subsequently removed by precision lapping/polishing.

Surface grinding processes can introduce highly deformed regions; this needs to be minimized to ensure that surfaces are representative of underlying microstructure. A stepwise grinding process with adequate cooling is recommended on both sides of the specimen to achieve the final thickness with the desired accuracy. One of the methods to verify the removal of disturbed layers is through microhardness measurements across the various layers from the specimen surface. For indentation-based techniques, the final sample preparation step is electropolishing, which removes the deformed surface layers and results in surface finish with very low roughness.

The focused ion beam (FIB) technique, which is commonly employed for fabrication of microelectromechanical system (MEMS) devices, is also gaining prominence for mechanical test sample preparation of both metallic and insulating materials with tolerances in the nanometer range. FIB uses a focused beam of ions, usually Ga+, to selectively remove material from a target material inside an evacuated chamber. The sputtering action enables precise machining of samples with accuracy of a few tens of nanometers with no mechanical predamage. However, for metallic specimens of sizes in the range of 0.5–5.0 mm, wire EDM followed by fine polishing is the most popular sample preparation method.

4.5 Uncertainty in Small Specimen Testing

The estimation of uncertainty is the key for standardization of any testing method because it is an index of reliability. "ISO Guide (2008) to the Expression of Uncertainty in Measurement" is widely used to estimate the uncertainty of testing methods (ISO 2008). If the measurand, which is the final property evaluated, can be expressed as y = f(x1,x2,...), the parameters (x1,x2,...) are defined as influence quantities. The sum of the uncertainties of all influence quantities is the uncertainty of the measurand. The quantity that influences the measurand is expressed by a probability density function (pdf) of the probable values of the quantity. Knowledge about a quantity is inferred from either repeated measurements (called type A evaluation), or from scientific judgment based on all the available information on the possible variability of the quantity (called type B evaluation). The mean of the pdf is taken as the best estimate of the value of the quantity. The standard deviation of the pdf is taken as the standard uncertainty in determining the value of the quantity. All identified standard uncertainty components, whether evaluated by type A or type B methods, are combined to produce an overall value of uncertainty known as the combined standard uncertainty associated with the result of the measurement (Adams 2002).

Though uncertainty estimation for conventional tests such as tension (Loveday 1999), hardness, etc. has been reported, similar efforts for mechanical tests with miniaturized specimens are limited. Jeon et al. (2009) performed uncertainty estimation for the tensile properties estimated from ball-indentation testing and compared results with those reported for uniaxial tensile test. Usually, uncertainty is determined from the experimental data using the mathematical relationship between the measurand and the influence quantities. However, in ball-indentation testing, Jeon considered the strength coefficient, work hardening, and yield strain as experimental data and assumed that they contain all uncertainty of experimental and calculation procedures. Using the same procedures of uncertainty estimation for conventional test methods, the average uncertainties for YS and UTS from round-robin spherical indentation experiments of five laboratories were estimated as 18.9% and 9.8%, respectively. This was seen to be higher than those reported for standard tensile tests, which generally are less than 5%. The higher uncertainty is partly attributed to the small size of the test specimen in indentation compared to the tensile specimen. The influence of local property variations and inhomogeneities in small volumes has larger influence on acquired data, while their influence is averaged out in large size specimens (several millimeters). The key to reduce uncertainty in small specimen testing is the optimization of experimental methods and the analysis procedures.

4.6 Round-Robin Exercises

Optimizing the test procedures for miniature specimen tests is an important requirement for standardization of the technique to ensure reproducible and accurate results that can be reliably correlated with data generated from full-size samples. Development of a standard for each of the small specimen techniques is an essential prerequisite for making this technology acceptable to the material testing community, designers, plant operators, and regulatory authorities. Toward this end, round-robin testing exercises in specific techniques within a country and across various countries have been in vogue to evolve a common code of practice (CoP) leading to an acceptable standard. The subject matter has been the focus of discussions in the ASTM series of conferences on small specimen test techniques held every four years since 1986 and international conferences on small specimen test techniques (SSTTs) conducted by the European group since 2010. Table 4.1 summarizes some of national and international round-robin programs in the area of small specimen testing. The details of the programs are described in van Walle et al. (2002), Odette et al. (2002), Takahashi et al. (1988), Xu et al. (2010), Matocha et al. (2012), and Hurst and Matocha (2015).

TABLE 4.1

Some National and International Programs in the Area of Small Specimen Mechanical Testing

Round-Robin Programs/Code of Practice/Standard	
Techniques	**Agency/Country**
Ball indentation, subsize/standard CVN, miniature disk compact tension, miniature fatigue/fracture, miniature tensile[a]	ASTM committee 10.02, 1994
Subsize Charpy testing[b]	ESIS TC 5 subcommittee
Reconstitution of Charpy specimens[c]	EC project—RESQUE
Deformation and fracture minibeam[d]	Japan–United States collaboration
Recommended practice for small punch of metallic materials[e]	Japan, Japan Atomic Energy Research Institute (JAERI)
Small punch for tensile, fracture, and creep behavior	EC CEN WS 21 (CWA 15627)
Small punch creep and miniature creep tests	Japan, Society for Material Science in Japan (JSMS)
Small punch[f]	China (China Special Equipment Inspection and Research Institute, East China University of Science and Technology, Nanjing University of Science)
Ball indentation	India, Department of Atomic Energy (DAE)

[a] Rosinski, S. T. and Corwin, W. R. 1998. In *Small specimen test techniques*, ASTM STP 1329, eds. W. R. Corwin, S. T. Rosinski, and E. V. Walle, 3–14. West Conshohocken, PA.

[b] Lucon, E. 1998. In *Small specimen test techniques*, ASTM STP 1329, eds. W. R. Corwin, S. T. Rosinski, and E. V. Walle, 15–31. West Conshohocken, PA.

[c] van Walle, E. et al. 2002. In *Small specimen test techniques*, STP 1418, ed. M. A. Sokolov et al., 409–425, West Conshohocken, PA: ASTM International.

[d] Odette, G. R. et al. 2002. *Journal of Nuclear Materials* 307–311:1643–1648.

[e] Takahashi, H. et al. 1988. Standardization of SP test (in Japanese). JAERI-M 88-172, JAERI, Ibaraki, Japan.

[f] Xu, T. et al. 2010. Study on standardization of small punch test. (1) General requirements. *Pressure Vessel Technology* 2010-07.

4.6.1 The European Code of Practice for Small Punch (SP) Testing

Given the state of knowledge of the SP test and the need to develop a set of guidelines to encourage uniformity in testing, CEN Workshop 21 on "Small Punch Test Method for Metallic Materials" was established in 2004. With the consensus of 32 participating organizations, CEN workshop agreement (CWA) 15627 "Small Punch Test Method for Metallic Materials" was issued in 2007. It comprises two parts: Part A: "A Code of Practice for Small Punch Creep Testing" and Part B: "A Code of Practice for Small Punch Testing for Tensile and Fracture Behavior." Annex B1, "Derivation of Tensile and Fracture Material Properties," describes the methods for estimation of yield and tensile strength, ductile–brittle transition temperature (DBTT), and fracture toughness of the metallic materials from SP test records. Annex B2, "Guidance on Relevant Technological Issues: Specimens Sampling from Components," describes potential applications of the SP test. The main elements of the code cover the apparatus, the test specimen preparation, test temperature considerations, test procedure, post-test examination and the approaches to the derivation of yield strength (YS), ultimate tensile strength (UTS), fracture appearance transition temperature (FATT), and fracture toughness from SP tests results.

Numerous investigations have been carried out employing SP tests using the CWA 15627 and presented in the biannual small sample test techniques symposia held since 2010. In one such study, using CWA 15627, an exercise on SP testing was carried out between Material & Metallurgical Research Ltd (MMRL), the Czech Republic; the Indira Gandhi Center of Atomic Research (IGCAR), India; and the University of Cantabria, Spain, for the determination of transition temperature T_{SP} on a circumferential welded joint of an outlet superheater header manufactured of forged tube and made of P22 steel (Matocha et al. 2012). The SP tests were carried out by all the participants on disk test specimens of 8 mm in diameter and 0.5 mm in thickness at temperatures ranging from 80 to 300 K using identical machines (screwdriven UTM Labortech make) installed in the respective laboratories. Figure 4.3 shows that the temperature dependence of fracture energy E_{SP} determined in each laboratory was in close agreement. It was also inferred that change of punch

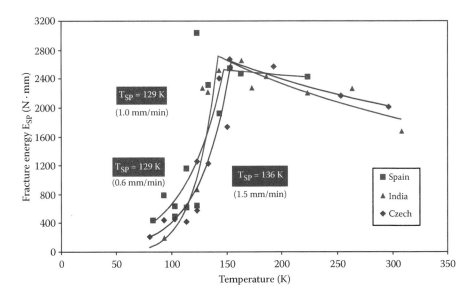

FIGURE 4.3
Comparison of the temperature dependence of fracture energy E_{SP} determined by three laboratories.

tip radius from 1.25 to 1.0 mm does not affect the temperature dependence of fracture energy in the transition region (Figure 4.4) because the load-punch displacement curves are modified by punch tip radius only at temperatures at which the ductile fracture is the dominant fracture mode (Figure 4.5).

Considerable insights have been obtained by researchers on applying the CoP to the industrial components. The orientation of SP test specimens was found to significantly affect the temperature dependence of fracture energy (T_{SP}) and empirical correlation with FATT, as seen in Figure 4.6. It was also seen that, in tough materials, the temperature dependence of fracture energy in the transition area was steep and the procedure recommended in the CWA for the determination of T_{SP} could lead to significant errors. Use of the specimen with a notch shifts transition temperature T_{SP} significantly to higher temperatures and the temperature dependence of fracture energy is also less steep. Based on these observations, Hurst and Matocha (2015) have outlined suggestions for improvement of the SP code and roadmap for prospective conversion into European standard. These include

- Use of the offset method (offset displacement of 0.1 mm) instead of the two tangent method for determination of F_e (yield load), resulting in significant lowering of scatter band of the empirical correlation

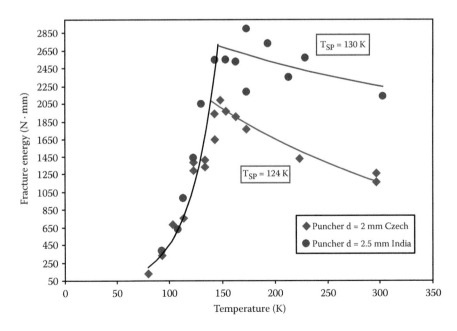

FIGURE 4.4
The effect of punch tip diameter (d = 2 mm, d = 2.5 mm) on temperature dependence of fracture energy.

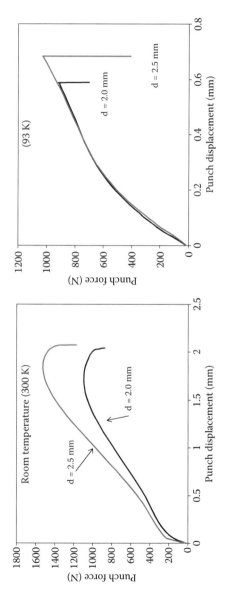

FIGURE 4.5
The effect of punch tip radius on load-punch displacement curve determined at room temperature and at 93 K.

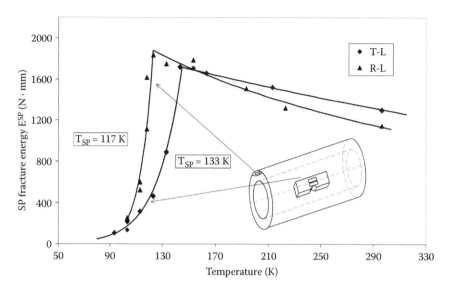

FIGURE 4.6
Effect of orientation of test disk on T_{SP} curves of 14MoV6-3 service exposed pipe steel.

- Use of force at first load drop for analysis when multiple load drops are observed in a load-displacement curve
- Incorporation of optimized notched specimen in the code
- Further studies on orientation of specimens in SP tests and the relationship with conventional Charpy V-notch (CVN) tests for industrial components

4.6.2 Round-Robin Experiments of Ball Indentation

A round-robin exercise on usage of ball indentation (BI) technique to predict strength properties of three different materials (carbon manganese steel [SA 333 Gr.6], stainless steel [SA 312 type 304LN], and zirconium alloy [Zr-2.5Nb]) used in Indian nuclear reactors was undertaken during 2008–2010. The main objectives of this round-robin exercise were

1. Quantification of variation in prediction of strength properties by participants
2. Assessment of extent to which effects of ageing on strength properties can be predicted

Five groups from the Department of Atomic Energy (DAE) of India participated in this round-robin exercise. Though all the participants used different test apparatus for carrying out ball indentation, common methodology

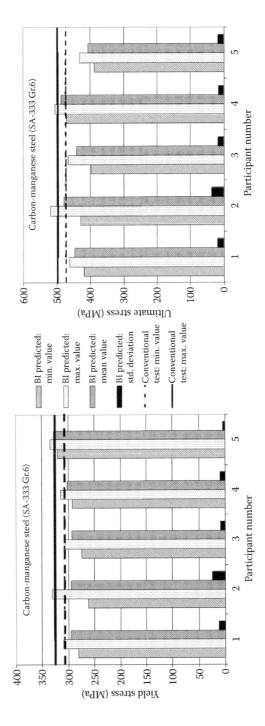

FIGURE 4.7
Comparison of ball indentation predicted YS and UTS by different participants and conventional test results.

similar to the one stated by Haggag was adopted for evaluating properties from load versus indentation depth curves.

A plot showing the comparison of BI predicted YS and UTS with their conventional equivalents of SA-333 steel for the five different participants is shown in Figure 4.7. The coefficient of variation (COV, defined as ratio of standard deviation to mean) in participants' results vary as follows:

- 1.3% to 8.0% for YS
- 3.2% to 7.7% for UTS
- 4.6% to 8.3% for K
- 5.2% to 21.2% for n

The overall COV in predicted YS and UTS is 5.14% and 7.68%, respectively, while the overall COV in predicted value of n is 26.93% and of K is 11.91%. The variation in values of YS and UTS is found to be much lower than that in n and K.

The various factors that could lead to differences between BI results and conventional tensile tests and the scatter among the various participants have been analyzed (Bhasin et al. 2015). In the BI process, the UTS prediction was based on n and K values, which in turn were evaluated from curve fit of stress-plastic strain pairs at each unloading stage. The scatter band in n and K basically arose due to choice and extent of best curve fit. The stress–strain behavior of most materials did not exhibit classical power law fit; that is, log (true stress) versus log (true plastic strain) was nonlinear. There was considerable scatter band in n and K; however, they were negatively correlated and hence the overall scatter band in predicted UTS was found to be significantly less. Another contributing factor to differences in results among participants was the extent of unloading and differences in linearization of unloading slope. It was seen that the nonlinearity in the lower part of the unloading curve leads to inconsistency in prediction of plastic indentation diameter (d_p), which in turn affects prediction of n, K, and UTS. It was, however, found that the BI technique is sensitive to the change in mechanical properties due to various types of treatment induced in these materials, while there were interparticipant variations in prediction of magnitude of change.

4.7 Concluding Remarks

In this chapter the various critical issues in small specimen testing were presented. The role of specimen size in deformation/fracture and the transportability between different specimen sizes are to be considered for effective use of small specimen test results. The uniformity in the test procedures and data interpretation methods is the need of the hour to improve and demonstrate a high level of accuracy and reproducibility of test results.

References

Adams, T. M. 2002. *A2LA guide for the estimation of measurement uncertainty in testing.* Frederick, MD: American Association for Laboratory Accreditation, Chap. 3.

Bhasin, V., Sharma, K., Singh, P. K., Vaze, K. K., Ghosh, A. K., Madhusoodan, K., Rupani, B. et al. Round-robin exercise on ball indentation technique in India: Indian nuclear reactor materials, nuclear engineering and design (accepted).

CEN Workshop Agreement CWA 15627:2007. 2007. Small punch test method for metallic materials. Brussels.

Efstathiou, C., Sehitoglu, H., and Lambros, J. 2010. Multiscale strain measurements of plastically deforming polycrystalline titanium: Role of deformation heterogeneities. *International Journal of Plasticity* 26 (1): 93–106.

Guduru, R. K., Nagasekhar, A. V., Scattergood, R. O., Koch, C. C., and Murty, K. L. 2007. Thickness and clearance effects in shear punch testing. *Advanced Engineering Materials* 9:157–159.

Hill, R. 1971. *The mathematical theory of plasticity.* Oxford, England: Clarendon Press.

Hurst, R. C., and Matocha, K. 2015. A renaissance in the use of the small punch testing technique. *Transactions SMiRT-23*, Manchester, United Kingdom, August 10–14.

Igata, N., Miyahara, K., Uda, T., and Asada, S. 1986. Effects of specimen thickness and grain size on the mechanical properties of types 304 and 316 austenitic stainless steel. In *The use of small-scale specimens for testing irradiated material*, ASTM STP 888, eds. W. R. Corwin and G. E. Lucas, 161–170. Philadelphia: ASTM.

ISO. 2008. Uncertainty of measurement—Part 3: 2008. Guide to the expression of uncertainty in measurement. ISO/IEC guide 98–3, International Organization for Standardization, Geneva, Switzerland.

Janssen, P. J. M., de Keijser, Th. H., and Geers, M. G. D. 2006. An experimental assessment of grain size effects in the uniaxial straining of thin Al sheet with a few grains across the thickness. *Materials Science and Engineering A* 419:238–248.

Jeon, E. C., Park, J. S., Choi, D. S., Kim, K. H., and Kwon, D. 2009. A method for estimating uncertainty of indentation tensile properties in instrumented indentation test. *Journal of Engineering Materials and Technology* 131:031006-1–031006-6.

Loveday, M. S. 1999. Room temperature tensile testing: A method for estimating uncertainty of measurement. CMMT, National Physical Laboratory, Teddington, UK, http://midas.npl.co.uk/midas/content/mn048.html.

Lucas, G. E., Sheckherd, J. W., and Odette, G. R. 1986. Shear punch and microhardness tests for strength and ductility measurements. In *The use of small scale specimens for testing irradiated material*, ASTM STP 888, eds. W. R. Corwin and G. E. Lucas, 112–140. Philadelphia: ASTM.

Lucon, E. 1998. Subsize impact testing: CISE experience and the activity of the ESIS TC 5 subcommittee. In *Small specimen test techniques*, ASTM STP 1329, eds. W. R. Corwin, S. T. Rosinski, and E. V. Walle, 15–31. West Conshohocken, PA: ASTM International.

Matocha, K., Filip, M., Karthik, V., Kumar, R., and Tonti, A. 2012. Results of round robin test for determination of T_{sp} of P22 steel by small punch test. *Proceedings of II International Conference on Small Samples Test Techniques* (SSTT 2), 227-232, Ostrava, Czech Republic, ISBN 978-80-260-0079-2.

Nagamani, Jaya B., and Alam, Md Zafir. 2013. Small scale mechanical testing of materials. *Current Science* 105:1073–1099.

Odette, G. R., He, M., Gragg, D., Klingensmith, D., and Lucas, G. E. 2002. Some recent innovations in small specimen testing. *Journal of Nuclear Materials* 307–311:1643–1648.

Oviasuyi, R. O. 2012. Investigation of the use of micro-mechanical testing to analyze the mechanical anisotropy of the Zr–2.5%Nb pressure tube alloy. PhD thesis, The University of Western Ontario.

Ranganathan, S. I., and Ostoja-Starzewski, M. 2008. Scale-dependent homogenization of inelastic random polycrystals. *Journal of Applied Mechanics, Transactions ASME* 75 (5): 0510081–0510089.

Rathbun, H. J., Odette, G. R., He, M. Y., and Yamamoto, T. 2006. Influence of statistical and constraint loss size effects on cleavage fracture toughness in the transition— A model-based analysis. *Engineering Fracture Mechanics* 73:2723–2747.

Ren, Z. Y., and Zheng, Q. S. 2002. A quantitative study on minimum sizes of representative volume elements of cubic polycrystals-numerical experiments. *Journal of Mechanics and Physics of Solids* 50:881–893.

Ren, Z. Y., and Zheng, Q. S. 2004. Effects of grain sizes, shapes, and distribution on minimum sizes of representative volume elements of cubic polycrystals. *Mechanics of Materials* 36:1217–1229.

Rosinski, S. T., and Corwin, W. R. 1998. ASTM cross comparison exercise on determination of material properties through miniature sample testing. In *Small specimen test techniques*, ASTM STP 1329, eds. W. R. Corwin, S. T. Rosinski, and E. V. Walle, 3–14. West Conshohocken, PA: ASTM International.

Simonovski, I., and Cizelj, L. 2007. Representative volume element size of a polycrystalline aggregate with embedded short crack. In *Proceedings of International Conference on Nuclear Energy for New Europe*, 0906.1–0906.8. Slovenia.

Song, M., Guan, K., Qin, W., Szpunar, J. A., and Chen, J. 2014. Size effect criteria on the small punch test for AISI 316L austenitic stainless steel. *Materials Science & Engineering A* 606:346–353.

Spary, I. J., Bushby, A. J., and Jennett, N. M. 2006. On the indentation size effect in spherical indentation. *Philosophical Magazine* 86 (33–35): 5581–5593.

Swadener, J. G., George, E. P., and Pharr, G. M. 2002. The correlation of the indentation size effect measured with indenters of various shapes. *Journal of the Mechanics and Physics of Solids* 50:681–694.

Tabor, D. 1954. The physical meaning of indentation hardness. *Sheet Metal Industry* 31:749–763.

Takahashi, H. et al. 1988. Standardization of SP test (in Japanese). JAERI-M 88-172, JAERI, Ibaraki, Japan.

Toloczko, M. B., Yokokura, Y., Abe, K., Hamilton, M. L., Garner, F. A., and Kurtz, R. J. 2002. The effect of specimen thickness and grain size on mechanical properties obtained from the shear punch test. In *Small specimen test techniques*, ASTM STP 1418, ed. M. A. Sokolov, J. D. Landes, and G. E. Lucas, 371–379. West Conshohocken, PA: ASTM International.

van Walle, E., Scibetta, M., Valo, H., Viehrig, W., Richter, H., Atkins, T., Wootton, M. et al. 2002. RESQUE: Reconstitution techniques qualification and evaluation to study ageing phenomena of nuclear pressure vessel materials. In *Small specimen test techniques*, STP 1418, ed. M. A. Sokolov et al., 409–425, West Conshohocken, PA: ASTM International.

Xu, T., Shou, B., Guan, K., and Wang, Z. 2010. Study on standardization of small punch test. (1) General requirements. *Pressure Vessel Technology* 2010-07.

Yang, L., and Lu, L. 2013. The influence of sample thickness on the tensile properties of pure Cu with different grain sizes. *Scripta Materialia* 69:242–245.

Zhu, T. T., Bushby, A. J., and Dunstan, D. J. 2008. Materials mechanical size effects: A review. *Materials Technology* 23 (4):193–209.

5

Applications of Small Specimen Testing

5.1 Introduction

Small specimen mechanical testing has primarily evolved out of the needs of the nuclear industry to develop and characterize materials of fission and fusion reactor systems. Apart from the numerous applications related to characterization of irradiated nuclear components/materials, it has far-reaching applications in areas ranging from residual life assessment of service-exposed components, alloy development programs, weld joints, failed components, and coatings to biomaterials and the electronic industry. In this chapter, a flavor of the various applications is presented to the readers.

5.2 Condition Monitoring of Plant Components

A reliable integrity assessment of structural materials is an essential prerequisite for safe and economic operation of an ageing plant component. The decision by plant operators to run or repair or replace a plant component is based on the assessment of (1) service loads and stresses and environmental conditions, (2) flaws in the component based on periodic nondestructive evaluation, and (3) material properties in the current state.

Extracting bulky samples from the component for tensile, fracture toughness, or creep properties is seldom done as this can cause considerable damage, resulting in forced outage, and can often require cumbersome and undesirable repair actions to allow further operation. If the assessment of in-service mechanical properties can be performed either by using a small allowable volume of material sampled nondestructively or using in situ tests in a minimally invasive manner, then the component can be continued in service with easy or no repair at all.

The electric power industry, both fossil fired and nuclear, and the petrochemical industry have found extensive application of small specimen test technology for life assessment of their components. The high-temperature components of power plants are designed for a definite lifetime, which is exhausted due to damaging mechanisms involving creep, fatigue, corrosion, and their combined influence. For components in the nuclear industry, the additional material degradation mechanisms include irradiation embrittlement, and irradiation-related swelling and creep.

5.2.1 Unique Challenges in the Nuclear Industry and Advantages of Small Specimens

The periodic assessment of material degradation of in-service nuclear components is routinely performed through (1) surveillance programs where prefabricated specimens are irradiated at specific locations in the nuclear reactor, and (2) specimens extracted from the irradiated component (such as the pressure vessel, pressure tube, or wrapper tube).

In the case of surveillance or material irradiation programs, standard bulk specimens not only consume the available reactor space during irradiation, but are also subjected to gradients in neutron flux and temperature. Small sized specimens can be conveniently accommodated in irradiation devices that fit in the available reactor space and thereby permit efficient use of material for irradiation experiments. Use of small specimens has an added advantage of uniform irradiation conditions, resulting in a homogenous microstructure within the irradiated volume.

The most challenging aspect of examining an in-service nuclear component is the induced radioactivity in the fuel, coolant and structurals caused by the neutrons in the reactor core. Due to high induced radioactivity, handling and testing of irradiated nuclear components is highly challenging and must necessarily be carried out remotely using robotic tools and with adequate shielding for protecting the operator from the gamma radiation (Kasiviswanathan 2001).

In the case of removable and/or replaceable components such as the fuel elements or assemblies, the mechanical property evaluation of irradiated materials is generally performed by transferring the component to shielded enclosures called the hot cells. For bulky and permanent structures such as the reactor pressure vessel, pipings, core shroud, etc., the sampling is performed at the site and the extracted sample is subsequently transferred to shielded enclosures for testing. Figure 5.1 shows the hot-cell facility at IGCAR, India, and a remotely operated tensile test machine housed inside the hot cell. The entire sequence of remote mechanical testing starting from machining the specimen, sample preparation, metrology, loading, gripping, and controlling the test and post-test handling is carried out remotely using master slave manipulators.

With miniaturized specimens, the gamma doses are significantly reduced to allow testing with minimal shielding, thus reducing cost and time. For specimen sizes of typically, say, 10 mm × 10 mm × 0.5 mm thick or less, gamma dose levels of neutron irradiated steels normally permit handling outside the hot cell with localized lead shielding around the test machine.

FIGURE 5.1
(a) The hot-cell facility (at IGCAR, India) for handling and testing irradiated components and (b) a remotely operated tensile test machine housed in a hot cell.

5.3 Sampling Techniques

The use of miniature specimens for condition monitoring of ageing plant components is closely tied with the ability to extract small size samples from a component or to employ portable field equipments for tests such as ball indentation directly on the component.

An in situ sampling technique is commonly employed to scoop material from operating components in a manner likened to an ice-cream scoop. Miniature specimens can be prepared from the scooped sample and the same can be tested for determining the mechanical properties of the operating component. Two types of scoop sampler are generally used: mechanical cutter systems and electrodischarge machining (EDM) systems. In the mechanical cutter type, the sample is obtained by spinning the cutter about its axis of symmetry, while slowly advancing a bit about a perpendicular axis to feed the cutter into the parent material. Figure 5.2 shows a miniature

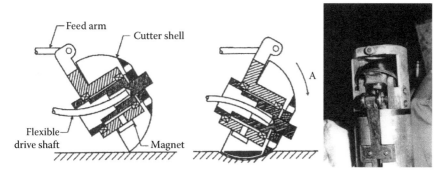

FIGURE 5.2
A scoop cutter for miniature sample removal patented by Exponent, Inc., of the United States.

sample removal system, patented by Exponent, Inc., of the United States (Parker, McMinn, and Foulds 1989). The cup-shaped sampler is coated with cubic boron nitride abrasive for grinding operation. While the typical scoop sample is about 25 mm in diameter and 3–4 mm in thickness, the actual shape and geometry depend upon the component and depth of cut setting. Figure 5.3 shows an example of an underwater sampler and a nuclear reactor pressure vessel sampler developed by Rolls Royce based on the SSam-2 (surface sampling system) technology. A remotely operable sampling module (Figure 5.4a and b) developed at BARC, India, has been deployed for scooping boat samples from underwater core components of the power reactors for life management studies (Kumar et al. 2014). The typical size of the boat sample (Figure 5.4c) scooped using this equipment is about 25 mm (width) × 40 mm (height) × 3 mm (thickness).

Extraction of small specimens from replaceable nuclear core components such as fuel assembly is carried out after transferring them from the reactor to shielded enclosures (hot cells). Figure 5.5 shows a hollow milling tool-based system employed in the hot cells for extraction of disk samples from irradiated AISI 316 stainless steel wrapper tubes of fast breeder test reactor (FBTR) for multiple examinations such as swelling, shear punch test, and electron microscopy.

An EDM-based sampling system employs a consumable electrode that removes material via electric spark erosion with an appropriate cooling medium. Unlike the traditional mechanical boat sampler, the surface damage in an EDM sampler is restricted to a few microns, resulting in minimal residual stresses generated in material removal. An EDM sampler facilitates a relatively large boat sample removal and a cutting profile (depth vs. surface area) that can be tailored to specific sampling needs. An EDM-based sampling system with special CuW alloy electrodes shaped in the form of a ring sector has been developed by CESI, Italy, for use in electric

(a)

(b)

FIGURE 5.3

(a) An underwater sampler and (b) a nuclear reactor pressure vessel sampler developed by Rolls Royce based on the SSam-2.

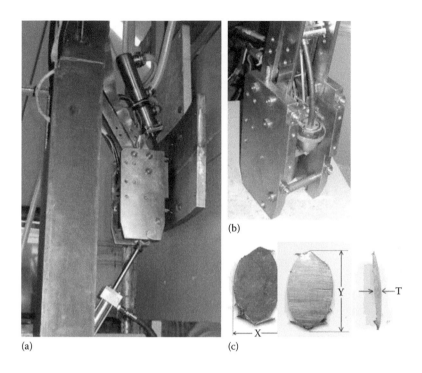

FIGURE 5.4

(a,b) Specimen sampling module positioned for scooping boat-shaped coupon from reactor core shroud, developed by BARC, India. (c) Geometry of boat sample 25 (X) × 40 mm (Y) × 3 mm (T).

FIGURE 5.5

Specimen extraction system using a hollow end mill tool employed in hot cells of IGCAR, India, for extracting small disk specimens from thin-walled irradiated components.

(a)

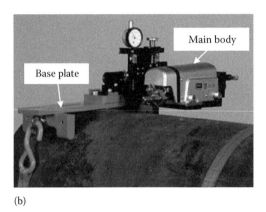

(b)

FIGURE 5.6
Images of electric discharge sampling equipment (EDSE) of ETD, UK showing (a) the EDSE head, the removed sample and (b) the main body and base plate.

power and process industry applications. An EDM sampler (Figure 5.6) has been developed by European Technology Development (ETD), UK, where the standard specimen thickness can be varied between 1 and 5 mm (other specimen dimensions being 20 × 25 mm) and the sampler has capability to extract specimens from the inside surface of pressure vessels (Kieran et al. 2012).

5.3.1 Considerations in Sample Removal

In addition to the approvals of equipment owners, operators and relevant regulatory authorities for sample removal from in-service components, there are several considerations in the selection of location and method of sampling:

- Property of interest and its relation to the damage and life-controlling mechanism operating in the component
- Stress intensification due to change in near surface geometry affecting the integrity of the component

- Damage to surface from heating and residual stresses
- Suitability of the surface (after sample removal) for future inspections by nondestructive evaluation

5.4 Field Equipment for In Situ Testing

Field instrumented systems based on instrumented ball- indentation testing capable of determining tensile properties at localized areas have been in vogue for the past 20 years. This is particularly useful because it avoids the need for sample removal.

Two leading manufacturers of instrumented indentation test equipment are Advanced Technology Corporation in the United States (Haggag 1999) and FRONTICS in Korea (Jang et al. 2003). Both these companies' equipment are based on multiple loading–unloading cycles of spherical indentation at a single location and evaluating the true stress–strain data based on the load-depth data and the morphology of the crater. For estimating mechanical properties on the inner side of a tubular component such as the irradiated pressure tube of the pressurized heavy water reactor (PHWR), an in situ property measurement system (IProMS) has been developed (Figure 5.7) and deployed by Bhabha Atomic Research Center (BARC), India (Chatterjee et al. 2014).

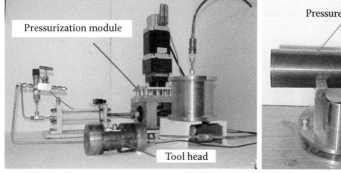
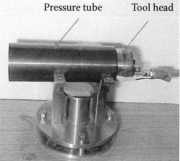

FIGURE 5.7
Ball-indentation-based IProMS with tool head, pressurization module, and IProMS placed inside pressure tube.

5.5 Residual Life Assessment

Creep, which is time-dependent deformation of materials subjected to mechanical load at high temperature, is a key factor not only in the design of components used in the power generation industry (fossil fuel plants and nuclear reactors), but also in the assessment of their remaining life.

Various nondestructive techniques have evolved for the evaluation of creep damage and remaining life assessment with respect to crack initiation. These include strain measurements, replication metallography, ultrasonic velocity measurements, hardness measurements, and electrical resistivity measurements (Kirihara et al. 1984). In Cr–Mo steels, where the principal damage under creep is thermal or strain-induced softening, hardness has been used as a measure of strength and as an index of creep life expended.

Vickers hardness testing has been used by many investigators to evaluate the reduction in the strength and as an index of creep life expended (Kimura et al. 1987). A methodology (Figure 5.8) was proposed by Gotoh (1985) for

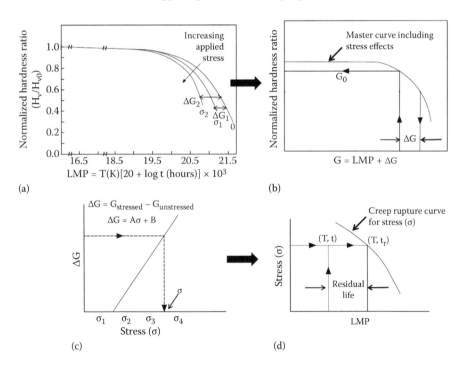

FIGURE 5.8
Schematic showing the methodology for estimating the expended creep life from hardness measurements. (a) Effect of stress on hardness ratio vs. LMP plot, (b) master plot of hardness ratio with modified LMP which includes both thermal aging and effects of stress, (c) plot of ΔG with the applied creep stress, and (d) standard creep rupture data from which creep life expended is determined for the stress obtained from (c).

predicting the remnant creep life of plant components based on the hardness data generated from laboratory-based creep tests at various stresses and temperatures. It was observed that application of stress accelerated the softening process and shifted the hardness to lower the Larson-Miller parameter values—LMP = T (20 + log t), where T is temperature in kelvins and t is time in hours—compared with the case of thermal aging (Figure 5.8a). The hardness changes (H/H_0) are related to time, temperature, and stress through a modified LMP defined as

$$G = G_0 + \Delta G = \log[T(20 + \log t)] + \Delta G \qquad (5.1)$$

where G_0 is the parameter describing the thermal softening behavior and ΔG is a parameter that incorporates the effect of stress. The parameter G thus includes the effects of both temperature and stress, and all the H/H_0 curves, regardless of the stress, can be normalized into a single master curve when plotted in terms of G (Figure 5.8b). The master plot of G versus H/H_0 is generated through laboratory experiments of measuring the hardness changes as a function of creep life expended at various stress levels. The parameter ΔG is determined as $G_{stressed} - G_{aged}$ when the creep-exposed material and thermal-aged condition show the same H/H_0. The relationship between ΔG and stress is expressed as

$$\Delta G = A\sigma + B \qquad (5.2)$$

On an actual plant component, hardness measurements are performed at unstressed and stressed locations. From the master plot of G versus H/H_0, the parameter ΔG is determined as $G_{stressed} - G_{unstressed}$, from which the local stresses are estimated (Figure 5.8c) using Equation 5.2. The time to rupture is obtained from the standard rupture data (Figure 5.8d) using the known values of stress and temperature from which the creep life expended is calculated. This method has been successfully applied for estimating the creep life consumption at the T-root corner of a rotor disk by Tanemera et al. (1988).

As an extension of this methodology, the ball-indentation method lends itself to practical applications of assessment of service degradation and aging management of power plant components with an added advantage of providing a design-based engineering parameter such as the ultimate tensile strength (UTS) or yield strength (YS). A study undertaken by Karthik et al. (2010) to quantify the progressive damage of mod 9Cr–1Mo steel under creep and thermal exposures consisted of a uniaxial creep test at 923K and terminated at predetermined fractions of creep life (Figure 5.9a). The head portion of the specimen corresponds to a thermally aged condition with microstructural changes caused by the effect of temperature alone, while microstructural changes in the gauge portion are induced by both temperature and stress. The strength and ductility changes of the different microstructural conditions have been evaluated using ball-indentation test techniques.

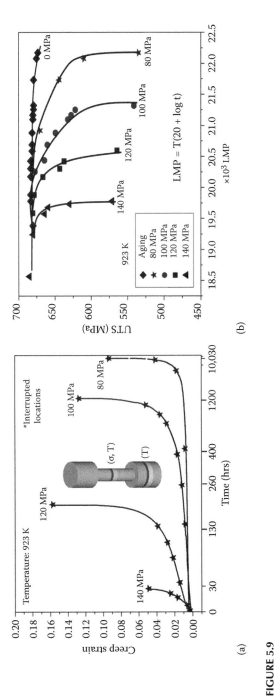

FIGURE 5.9

(a) Creep curves of conventional creep tests carried out at various stress levels (*indicates the instances where tests were terminated and specimen removed). (b) Plot of the UTS with LMP—T(20 + log t)—for the various stressed and unstressed exposures.

Combining the temperature and time variables through LMP, UTS is plotted against LMP in Figure 5.9b for various stressed and unstressed conditions; this has a physical form similar to that of the hardness profile. Thus, the ball-indentation technique can be suited for assessing the residual creep life of power plant components through the use of portable test equipment.

5.6 Properties of Weld Joints

The knowledge of mechanical properties of a weld joint is essential for understanding the behavior of the welded structures employed in high-performance applications. A welded joint consists of unaffected base metal, weld metal, and heat-affected zones (HAZs), which are metallurgically and chemically heterogeneous (Figure 5.10). This is due to the temperature gradients associated with the welding process as a function of distance from the fusion zone and the chemical gradients that evolve during the process. This in-homogeneity is most severe in dissimilar or multipass welds, resulting in gradients in mechanical properties across the individual zones of the weldments.

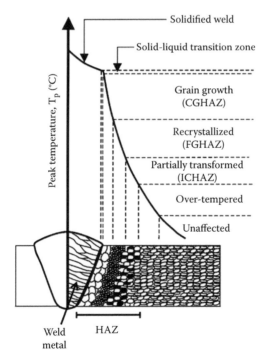

FIGURE 5.10
Schematic representation of peak temperatures as a function of distance from the fusion line and the resulting microstructures across a steel weldment.

Weld joints are subjected to standard destructive tests such as hardness, bend, and tensile for developing the welding procedure specification and assessing the suitability of a weld joint for a particular application. In such cases, the weld joint is assessed by considering it a single composite entity by testing a complete section of the weldment. For example, tensile properties of the weld joint are obtained in two ways: (1) taking specimens from a transverse direction of weld joint consisting of base metal-HAZ-Weld metal (Figure 5.11a), and (2) all weld metal specimens as shown in Figure 5.11b. Such methods often result in failure at base metal, yielding only information that the fusion zone and the HAZ strengths exceed those of the base metal.

However, when damage is concentrated in specific regions of the fusion zone or the HAZ—for example, in creep exposed joints of chrome–moly ferritic steels—it is desirable to have knowledge of the individual zone of the weld joint for better design of alloy, welding, and postweld heat treatment procedures and predictions of remaining life. It is practically impossible to obtain standard specimens cut from individual regions of the weldment for evaluating mechanical properties by conventional test techniques due to the narrowness of the various zones. One approach is to simulate these microstructures through special heat-treatment techniques and obtain sizeable volumes of material that would approximate the individual weld/HAZ microstructures for carrying out mechanical tests.

Alternatively, the use of disk specimen tests such as the small punch (SP), shear punch, and indentation-based techniques are promising for application in localized zones of weld joints. Hardness testing is the usual approach for delineating the properties of these various zones, but the information obtained is very limited. As an extension of this method, the ball-indentation technique, which is capable of providing tensile properties such as YS, UTS, and strain-hardening exponent, has been extensively employed for localized zones across a weld joint. The indentation method is well suited for the fusion zone (the weld metal), where the plastic zone and constraint are well contained within the zone and the properties estimated are representative of the

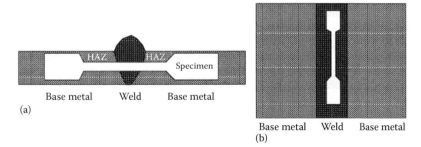

(a)

(b)

FIGURE 5.11
Schematic showing tensile specimen sampling from a weld joint. (a) Tensile specimen sampled across the weld joint consisting of base metal, weld metal and HAZ. (b) Tensile sample sampled longitudinally from the weld joint consisting the fusion zone only.

underlying microstructure. However, for the different microstructural regions spanning a few millimeter widths within the HAZ, the plastic zone around the indentation is seldom confined to individual microstructural zones and the estimated properties are more averaged out values of the adjacent layers. Indenter sizes of diameters less than 0.2 mm and down to nanometer ranges are commonly employed for probing the individual microstructures in narrow HAZs. As compared to indentation-based techniques, punch out tests such as shear punch and small punch (SP) are better suited for probing the HAZs as the specimen sizes are more or less the same as the width of the individual zones. However, the main challenge is identifying and extracting specimens with uniform microstructure from the individual zones.

Shear punch testing has been put to use by Karthik et al. (2002) for evaluating the strength and ductility changes across the HAZs of 2.25Cr–1Mo steel weld joints, which consist of coarse-grained bainite, fine-grained bainite, and intercritical microstructures. By testing small specimens extracted from the individual regions of the HAZ in the weld joint (Figure 5.12), the tensile property gradients across the HAZ have been evaluated and related to their respective microstructural features. The methodology was based on tensile-shear correlations derived for the strength and ductility parameters using different microstructural conditions of the steel. Similar studies have been carried out by Stewart et al. (2006) for measuring the variation in mechanical properties of welded pipe line of 4130 steel at room temperature and over a range of potential service temperatures down to 213 K. The results showed that the shear punch technique could predict the properties with a resolution intermediate between those of tensile testing and hardness testing, with the advantage of both strength and ductility measurement unlike hardness tests.

Gulcimen et al. (2013) have employed SP tests to determine the transition temperature T_{sp} of the individual zones of a P91 steel weldment such as weld metal, base metal, fine-grained HAZ, and coarse-grained HAZ using specimens extracted from the individual zones and testing in the temperature range from 298 to 77 K. It was observed that the SP energy has a higher scatter compared to the Charpy data. The behavior is attributed to the reduced specimen size, which restricts the number of weak links to be probed. However, the SP test could distinguish the different zones of steel weldment with respect to the lower temperature mechanical behavior.

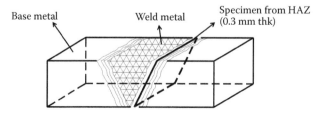

FIGURE 5.12
Schematic showing the sampling of a disk specimen from the HAZs of a weldment.

5.7 Coatings and Surface-Treated Components

Surface treatment is an importance process in the production of machine parts, generally employed to enhance the structural element strength and improve specific properties such as fatigue life, corrosion resistance, etc. Surface deposition or coatings are commonly applied for automotive parts, cutting tools, integrated circuits, and biomaterials and in high-temperature components of power-generating/chemical-processing plants.

Usually the surface layer (whose thickness ranges from few micrometers to fractions of millimeters) is characterized by microhardness, roughness, adhesion, and wear resistance. To obtain plastic stress–strain properties, the indentation technique is preferred because of its capability of deforming materials on a very small scale. The measurement of coating-only properties requires elimination of the substrate effects by making indentation within a critical penetration depth beyond which the substrate effects appear. A commonly held view is that the penetration depth should be less than 10% of the coating thickness to ensure that the plastic zone underneath the indenter does not extend to the coating–substrate interface. Extensive studies by Gamonpilas and Busso (2004) show that, for hard coating applied on a soft substrate, this limit comes down to 5% of the coating thickness to avoid substrate effects.

In one such application, ball-indentation test method was employed by Kucharski and Radziejewska (2003) for evaluating the plastic properties of laser alloyed layers (cobalt stellite and tungsten alloyed layers) on carbon steel. For a melted zone of 190 μm thickness identified through microhardness measurements, the maximum depth of penetration was limited to 9 μm using a 400 μm diameter spherical indenter. However, for very small coating thickness, incremental load during the multiple loading–unloading cycles becomes too small, leading to an increase in relative error in the force or depth measurement. In recent years, nanoindentation methods employing high-resolution instrumentation to continuously control and monitor the loads in the range of millinewtons to newtons and displacements in nanoscale have gained wider acceptance for quantitative assessment of thin surface depositions.

Typical high-temperature components where coatings play an important role are steam turbines, heat exchangers, supercritical boilers, superheaters, aero-gas turbines, industrial gas turbines, and diesel engines. The coatings provide protection against oxidation, corrosion, and mechanical stresses of the superalloy substrate. The fracture strain that a coating can tolerate changes as a function of temperature, and it is important to determine the temperature dependence and ductile-to-brittle transition temperature (DBTT) of the coating ductility to assess its performance. Many researchers have employed small punch (SP) test methods for such an assessment due to restrictive sizes of material available from a virgin as well as service-exposed component. Saunders, Banks, and Wright (2001) employed SP tests at elevated temperatures up to 1000°C to determine DBTT of coatings (Pt aluminized, Sicoat 2231,

MCrAlY) deposited on IN738, CMSX4, Rene80 substrates. They used finite element (FE) analysis to convert the vertical displacement measured at the center of the sample to in-plane tensile strain in the coating and employed online acoustic emission monitoring to detect the onset of cracking in SP tests. The criterion used to define DBTT in their study was the temperature at which the strain to first crack is at a certain level—say, 10%. In another study reported by Eskner and Sandstrom (2003), SP test was used for characterizing the DBTT of nickel aluminide coatings deposited on nickel base gas turbine blades. Disk samples 3 mm diameter and around 0.1 mm in thickness prepared from NiAl coating were SP tested at various temperatures from ambient to 860°C. Typical load displacement of SP tests at various temperatures (Figure 5.13) showed plastic deformation above about 800°C and brittle fracture for all test temperatures below it. The DBTT of the coating, defined from the plot of strain to fracture with test temperature was estimated as 760°C.

Shear punch and indentation-based techniques (Ganesh et al. 2008) have been employed for characterizing multimaterial components made by laser rapid manufacturing (LRM). LRM represents an additive manufacturing process involving layer-by-layer fabrication of a three-dimensional (3D) engineering component with a flexibility to tailor microstructure, chemical composition, and its gradient in the resultant component. A bimetallic tube of 25 mm inner diameter (ID) and 3.8 mm wall thickness fabricated with Stellite 21 on ID and type 316L SS on outer diameter (OD) by LRM was studied (Figure 5.14a and b). The load displacement of the ball-indentation test and resultant

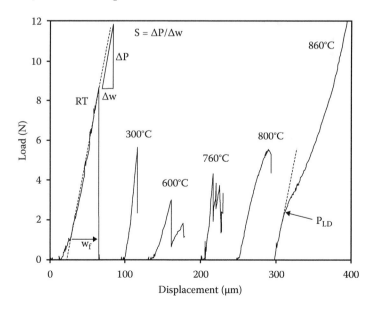

FIGURE 5.13
Load-displacement curves of SP tests on specimens from nickel alumide coatings at different temperatures exhibiting plastic deformation at temperatures above 800°C.

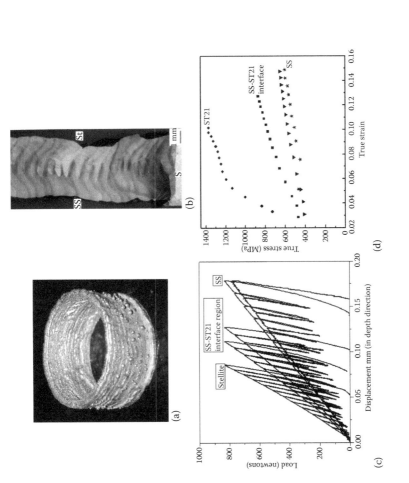

FIGURE 5.14

(a) A fabricated bimetallic tube (ID: ST21;OD: SS). (b) Photomacrograph of longitudinal cross sections of bimetallic tube. (c) Load depth and (d) flow curves obtained from different locations across the cross section of the bimetallic wall. SS: stainless steel 316L; SS–ST21: interface region of SS/Stellite 21; ST21: Stellite 21.

stress–strain data in the various layers as shown in Figure 5.14c and d confirm the viability of small specimen techniques for such applications.

5.8 Material Development Programs

Materials have been fundamental to the development of civilization. From the Bronze Age to the silicon-driven Information Age, civilization has defined itself and has advanced by mastering new materials. Advances in materials have driven our economic, social, and scientific progress and profoundly shaped our everyday lives.

Small specimen testing is of great advantage during the development phase of a new material processing. Often the quantities of new and exotic materials processed on a laboratory scale during the development phase would be insufficient for conventional specimen sizes and hence testing with small specimens becomes valuable for rapid screening and optimizing the processing routes.

The small specimen methodologies have found applications for characterizing prehistoric and precious materials as demonstrated by Lacalle et al. (2010). They have used SP tests for characterizing cast iron and bronze materials discovered in archaeological sites in Spain and also for screening gold of different purity contents by estimating both strength and toughness of gold as a function of alloying elements such as silver and copper.

5.8.1 Nanomaterials and Composites

Nanomaterials are materials where the sizes of the individual building blocks are less than 100 nm, at least in one dimension; they have properties dependent inherently on the small grain size. Conventional engineering materials are manufactured by starting from large pieces of material and producing the intended structure by mechanical or chemical methods (top-down approach). Nanomaterials are made through controlled manipulation of size and shape of materials at the nanometer scale (i.e., atoms or molecules as building blocks) through suitable chemical or physical processes (bottom-up approach), as shown in Figure 5.15.

Nanocrystalline (nc) metals and alloys are defined as those with an average grain size and range of grain sizes smaller than 100 nm. Such materials show superior yield and fracture strengths, decreased elongation and toughness, and superior resistance to wear and environmentally assisted damage compared to conventional grain size materials. The mechanisms of deformation and the properties of nc material not only depend on the average grain size, but are also strongly influenced by the grain size distribution and the grain boundary structure. As grain size approaches the size of dislocation in

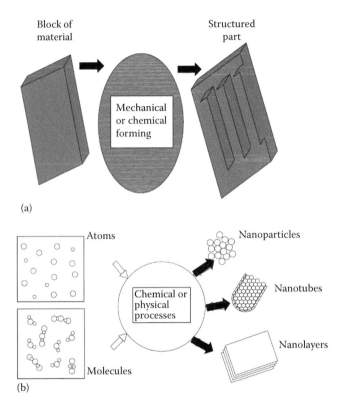

FIGURE 5.15
Schematic showing (a) the top-down approach for manufacturing conventional grain size materials and (b) the bottom-up approach for synthesizing nanograined materials.

nc materials, mechanical deformation is dominated by grain-boundary sliding and twinning processes.

The processing techniques currently available to produce nc materials are generally classified into the following four groups: mechanical alloying (including cryomilling) and compaction, severe plastic deformation, gas-phase condensation of particulates and consolidation, and electrodeposition. Evaluating the true mechanical behavior of nanomaterials is challenging due to the difficulties inherent in fabricating flaw-free, large-scale materials. Processing induced defects such as porosities or cracks and surface roughness has been found to have a significant influence on the measured mechanical properties, making it difficult to understand the intrinsic deformation characteristics of materials with nanometer grain size.

Synthesizing bulk nanocrystalline materials on laboratory scale for optimizing the process parameters and characterization often results in small quantities, which limits making large-scale specimens for conventional testing such as tension, compression, etc. Small specimen tests employing disk specimen geometry are very promising for such applications. Zhang et al. (2009) employed shear punch tests for assessing the mechanical behavior of bulk nanostructured Ag rods (average particle size of 30–50 nm) fabricated using dynamic compaction of spherical nano-Ag powders (99.9% purity). This study indicated a strong connection between surface finish and mechanical properties; samples exhibiting a smooth surface had a yield strength of 310 MPa, ultimate tensile strength of 320 MPa, and ductility of 15%, while the samples with a rougher surface displayed a brittle behavior and a reduction in strength of 35% (Figure 5.16). The shear punch results revealed that the surface finish or surface wear needs to be carefully controlled to avoid catastrophic failure in the use of bulk nanomaterials in structural applications.

Guduru et al. (2006) demonstrated the applicability of shear punch techniques for characterizing nanocrystalline Cu synthesized by an electrodeposition process. Subsize tensile (7 mm total length, 2 mm gauge length, 1 mm gauge width) and shear punch tests were employed by Guduru and colleagues to determine the mechanical properties of electrodeposited nc copper (grain size of 74 nm) and coarse-grained Cu (reference material for comparison) and also to determine the strain rate sensitivity and flow stress activation volume. The results of both subsize tensile and shear punch techniques revealed high strength, strain-hardening capacity, and moderate tensile ductility of nc Cu (compared to coarse-grained Cu). The study further revealed that the strain rate sensitivity increased and the activation volume decreased when the grain size of face centered cubic (FCC) Cu decreased to the nanometer range.

The ever increasing demand for newer, stronger, stiffer, and lighter weight materials in fields such as aerospace, transportation, and construction has led to the development of a newer class of materials called composites.

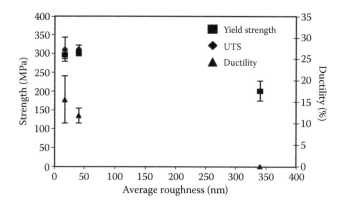

FIGURE 5.16
Comparison of strength and ductility of nanostructured Ag rods from shear punch testing as a function of surface finish.

Composites are made by combining two or more materials to give a unique combination of properties better than those of the individual components used alone. As opposed to metal alloys, each material retains its chemical, physical, and mechanical properties. The two constituents are normally the reinforcements (fibers or particulates or whiskers) and the matrix may be metal (MMC), polymer (PMC), or ceramic (CMC).

Tensile testing of composites and brittle materials is complicated by the need for perfect specimen alignment and a specimen surface free of flaws or machining grooves. Alternate techniques such as shear punch have the advantage of small and simple specimen configurations with reduced cost and hence are promising for research and development activities of composite materials. For different discontinuously reinforced metal matrix composites (such as Al reinforced with Al_2O_3, Ti–6Al–4V reinforced with TiC particles), the work of Wanjara, Drew, and Yue (2006) established that, like metallic alloys, the deformation and failure behavior of composites in shear punch testing resembles that in uniaxial tension and that corresponding properties could be linearly correlated. Zabihi, Toroghinejad, and Shafyei (2014) employed shear punch techniques for estimating the mechanical properties of Al/alumina composite produced using powder metallurgy and then processed by a hot rolling procedure. Their study established that mechanically milled pure Al showed higher strength compared to hot rolled pure Al strips and the shear strength increased by increasing the amount of alumina particles in the aluminum matrix. The investigators preferred shear punch over tensile tests due to the small size requirement for punch tests.

5.8.2 Metallic Glass

Metallic glasses, or amorphous metals, are novel engineering alloys that lack crystallinity and microstructural features such as grain and phase boundaries (Ashby and Greer 2006). The two most attractive properties of metallic glasses are high-yield strength far exceeding the strength values of crystalline metals and alloys and superior elastic strain limit: that is, the ability to retain original shape (memory) after undergoing very high loads and stress. This can be seen in the plot of strength (σ_y) with modulus (E) for a range of common materials (Figure 5.17). Metallic glass finds applications in specific products and high value-added metal components, such as consumer electronics frames and casings; in nuclear industries for making containers for nuclear waste disposal and magnets for fusion reactors; and in biomedical devices such as orthopedic screws, cardiovascular stents, and precision surgical instruments.

Bulk metallic glasses (BMGs) undergo highly localized, heterogeneous deformation by formation of shear bands. Under uniaxial tension tests, metallic glass fails in a brittle manner with unstable propagation of shear

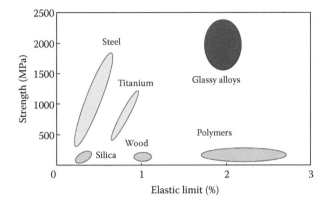

FIGURE 5.17
Figure showing amorphous metallic alloys combining higher strength of crystalline metal alloys with the elasticity of polymers.

bands. In contrast to crystalline alloys where plastic deformation is strongly influenced by the local shear stresses, without any pronounced effects of normal and hydrostatic stresses, the elastoplastic deformation response in BMG is influenced by the shear as well as the normal component (pressure) of the local stress, and possibly by the hydrostatic stress. Guduru (2006) made use of shear punch (ShP) tests for characterizing the shear deformation response of bulk metallic glasses based on the premise that shear-dominant loading in ShP test is more appropriate for determining the intrinsic (zero pressure) plastic flow properties. Guduru showed a good agreement between ultimate shear stress values (obtained from shear punch tests) with the shear failure stress from compression loading reported for the metallic glass Zr-5Ti-17.9Cu-14.6Ni-10Al. The ball-indentation-based data (triaxial compressive loading) showed a lack of pressure effect for this class of metallic glass. This validation provided the impetus to make use of shear punch for measuring an intrinsic (pressure independent) ultimate shear stress for other bulk metallic glasses, most of which show pressure sensitivity for both compression and tension testing.

5.8.3 Biomaterials

Ultrahigh molecular weight polyethylene (UHMWPE) orthopedic bearings are extensively used in total hip, knee, and shoulder arthroplasties. They contain between 40 and 200 g of material, and the designs typically have characteristic length scales of between 4 and 60 mm (Edidin 2009). The mechanical behavior of UHMWPE is dependent on its processing, sterilization, and cross-linking history and further evolves during shelf storage or after implantation into the human body.

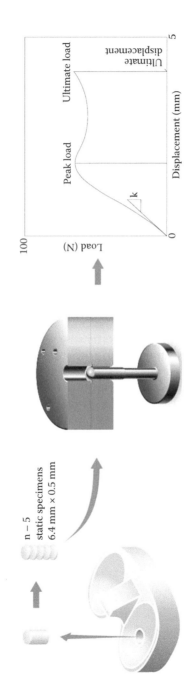

FIGURE 5.18
Schematic showing specimen preparation from UHMWPE component and representative load-displacement curve showing the various parameters.

Historically, mechanical tests of UHMWPE were performed in uniaxial tension using specimens such as those described in ASTM D648. Considering the small sizes of specimen required for small/shear punch tests, which enables testing multiple specimens across the thickness of the component, these test methods have been adapted (Kurtz et al. 1997) for characterizing the mechanical behavior of UHMWPE. This is of particular interest to study of oxidative degradation, wear-induced chain alignment, or any other mechanism that leads to a material inhomogeneity through the thickness of the bearing. Another motivation for use of punch tests for UHMWPE was the fact that the uniaxial test mode was found not to be predictive of the types of wear damage observed in vivo, which are generated by more complex, multiaxial loading conditions. Small punch (SP) tests have proven to be useful in the detection of the mechanical signature of oxidative degradation and of irradiation-induced cross linking (Kurtz et al. 1999). A schematic of the specimen extraction from an orthopedic component and the typical load-displacement curve of an SP test with the metrics is shown in Figure 5.18. Kurtz and colleagues further adapted the shear punch technique as a complementary tool for a better understanding of the wear resistance of UHMWPE, which was initially believed to be attributed to some kind of cross-shear behavior, but subsequently dismissed as misnomer. Both shear and SP tests have been very useful for understanding the relationship between the mechanics of UHMWPE to both adhesive/abrasive wear and mechanical breakdown of arthroplasty-bearing components.

Subsequently, a cyclic SP testing has also been developed to understand the fatigue mechanisms associated with the failure of the bearing components such as delamination and frank failure. In the method adopted by Villarraga et al. (2003), static SP tests were performed a constant loading rate of 200 N/s to determine the mean peak load (P_{max}), followed by fatigue punch tests to a maximum load of 0.55 to 0.90 P_{max}, depending on the number of cycles to failure expected or desired (Figure 5.19). Two metrics comparing the relative

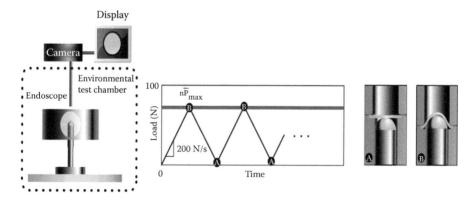

FIGURE 5.19
Schematic of fatigue punch methodology of Villarraga et al. (2013) for characterizing UHMWPE components.

fatigue resistance of a given material were the S–N curve and the hysteresis (relative energy retained during each cyclic excursion).

The shear punch technique has also been extended to dental restorative materials and has been successfully employed (Nomoto, Carricka, and McCabe 2001) to differentiate the mechanical strength of different restorative materials such as amalgam, composite resin, compomer, composite containing prereacted glass ionomer filler, resin modified glass ionomer, and polycarboxylate cement.

5.9 Electronic Industry

Extensive efforts have been undertaken in the last decade toward development of lead-free solders for applications in electronic packaging, considering the harmful effects of lead-containing solders. Alloys from the Au–Sn, Au–Ge, Zn–Al, Zn–Sn, Bi–Ag, and Sn–Sb systems are potential replacements for Pb–Sn solders for high-temperature soldering applications. Accurate mechanical properties and constitutive equations for solder materials are needed for use in mechanical design, reliability assessment, and process optimization. Fatigue of the solder joint is one of the predominant failure mechanisms in lead-free electronic assemblies exposed to thermal cycling. Most test methods are either based on uniaxial testing of miniature bulk solder specimens under tensile or compression mode. With solder joints as small as 0.5×0.5 mm, the actual mechanical behavior of the solder joint at the device level is seldom studied by miniature tensile/compression due to size limitations. Currently, the nanoindentation technique is gaining prominence in probing the mechanical properties of such small sizes.

Long before nanoindentation test methods were in vogue, Murty and Haggag (1996) employed the ball-indentation technique to study the mechanical behavior of Sn–5%Sb solder material. The strain-rate sensitivity (SRS) of Sn–5%Sb was determined by conducting ball indentation at varying test speeds compared with tensile and creep tests. From the dependence of the stress on strain rate, the activation area for deformation and controlling mechanisms (e.g., climb of edge dislocations and motion of jogged screw dislocations) was derived.

5.10 Concluding Remarks

Though the conventional test methods have been the mainstay of worldwide design codes for the design and manufacture of all components, the assessment of material properties of in-service structural components either using small, scooped-out specimens or by in situ techniques is promising as it is minimally invasive. The range of applications of small specimen testing with specimen sizes in the 0.5–5.0 mm range highlighted in this chapter for mechanical property characterization underpins its relevance to current technological issues and the associated challenges it presents to the measurement of stress and strain in small volumes. This specialized subject topic is of interest not only to industries associated with power generation (nuclear and conventional), aerospace, defense, oil, and gas, but also to materials scientists dealing with materials development and deformation studies. Though this technology can be claimed to have reached a state of maturity through extensive studies of numerous researchers in the last three decades, the future lies in formulating acceptable standards and transferring the methodologies to provide real-time, in situ measurement of properties on components in the field.

References

Ashby, M. F., and Greer, A. L. 2006. Metallic glasses as structural materials. *Scripta Materialia* 54:321–326.

Chatterjee, S., Panwar, S., Madhusoodanan, K., Vijayan, P. K., Mallik, G. K., and Alur, V. D. 2014. Measurement of mechanical properties of a PHWR operated pressure tube using an in-house developed in situ. Property measurement system (IProMS). *BARC Newsletter* 339:11–17.

Edidin, A. A. 2009. Development and application of the small punch test to UHMWPE. In *UHMWPE biomaterials handbook*. Burlington, MA: Academic Press, Inc.

Eskner, M., and Sandstrom, R. 2003. Measurement of the ductile-to-brittle transition temperature in a nickel aluminide coating by a miniaturized disc bending test technique. *Surface and Coatings Technology* 165:71–80.

Gamonpilas, C., and Busso, E. 2004. On the effect of substrate properties on the indentation behavior of coated systems. *Materials Science and Engineering A* 380:52–61.

Ganesh, P., Karthik, V., Kaul, R., Paul, C. P., Tiwari, P., Mishra, S. K., Prem Singh, C. H. et al. 2008. Fabrication of multimaterial components by laser rapid manufacturing and their characterization. In *Proceedings of 17th International Conference on Processing and Fabrication of Advanced Materials*, eds. N. Bhatnagar and T. S. Srivatsan. I. K. International Publication House Ltd. 1:AC-127.

Gotoh T. 1985. Study of residual creep life estimation using nondestructive material property test. *Mitsubishi Technical Bulletin*, no. 169.

Guduru R. K. 2006. Mechanical behavior of nanocrystalline materials and application of shear punch test. PhD thesis , North Carolina State University, Raleigh, NC, USA.

Guduru, R. K., Darling, K. A., Scattergood, R. O., Koch, C. C., Murty, K. L., Bakkal, M., and Shih, A. J. 2006. Shear punch tests for a bulk metallic glass. *Intermetallics* 14:1411–1416.

Gulcimen, B., Durmus, A., Ülkü, S., Hurst, R. C., Turba, K., and Hähner, P. 2013. Mechanical characterization of a P91 weldment by means of small punch fracture testing. *International Journal of Pressure Vessels and Piping* 105–106:28–35.

Haggag, F. M. 1999. Nondestructive determination of yield strength and stress–strain curves of in-service transmission pipelines using innovative stress–strain microprobe TM technology. ATC/DOT/990901. US Dept. of Transportation, Washington, DC.

Jang, J. I., Choi, Y., Lee, Y. H., Lee, J. S., Kwon, D., Park, J., Gao, M., and Kania, R. 2003. Instrumented indentation technique to measure flow properties: A novel way to enhance the accuracy of integrity assessment. In *ASME 2003 22nd International Conference on Offshore Mechanics and Arctic Engineering.* Volume 3: Materials technology, ocean engineering, polar and arctic sciences and technology workshops. Cancun, Mexico.

Karthik, V., Laha, K., Chandravathi, K. S., Parameswaran, P., Kasiviswanathan, K. V., and Raj, B. 2010. Ball-indentation test technique for evaluating thermal and creep damage of modified 9Cr–1Mo steel. *Transactions of the Indian Institute of Metals* 63(2–3):431–436.

Karthik, V., Laha, K., Kasiviswanathan, K. V., and Raj, B. 2002. Determination of mechanical property gradients in heat-affected zones of ferritic steel weldments by shear-punch tests In *Small specimen test techniques,* ASTM STP 1418, eds. M. A. Sokolov, J. D. Landes, and G. E. Lucas, 380–405. ASTM, Philadelphia.

Kasiviswanathan, K. V. 2001. Hot cells, glove boxes and shielded facilities. In *Encyclopedia of materials science and engineering,* eds. K. H. Jurgen Buschow, R. W. Cahn, M. C. Flemings, B. Ilschner, E. J. Kramer, and S. Mahajan, 4:3830. Elsevier, New York.

Kieran, C., Shibli. A., Fernandes. J., and Smith. C. 2012. Experience with electrical discharge sampling equipment in sample removal for miniature specimen testing and material quality checks. In *Proceedings of II International Conference on Small Samples Test Techniques (SSTT 2).* Determination of mechanical properties of materials by small punch and other miniature testing techniques. ISBN 978-80-260-0079-2.309-318. Ostrava, Czech Republic.

Kimura, K. et al. 1987. Life assessment and diagnostic system for steam turbine components. In *Life extension and assessment of fossil power plants,* eds. R. B. Dooley and R. Viswanathan, 677–685. EPRI CS 5208. Electric Power Research Institute. Palo Alto, CA.

Kirihara, S., Shiga, M., Sukekawa, M., Yoshioka, T., and Asano, C. 1984. Fundamental study of nondestructive detection of creep damage for low alloy steel. *Journal of the Society of Materials Science* 33:1097–1102. Japan.

Kucharski, S., and Radziejewska, J. 2003. Microindentation test for charcaterization of plastic properties of laser alloyed layers. *Journal of Testing and Evaluation* 31(2):106–115.

Kumar, K., Madhusoodanan K., Singh, R. N., Chakravartty, J. K., Dutta, B. K., and Sinha, R. K. 2014. Development and validation of in situ boat sampling technique for mechanical property evaluation and life management of TAPS core shroud. *International Journal of Engineering Research & Technology (IJERT)* 3(7):1702–1708.

Kurtz, S. M., Foulds, J. R., Jewett, C. W., Srivastav, S., and Edidin, A. A. 1997. Validation of a small punch testing technique to characterize the mechanical behavior of ultra-high molecular weight polyethylene. *Biomaterials* 18(24):1659–1663.

Kurtz, S. M., Jewett, C. W., Foulds, J. R., and Edidin, A. A. 1999. A miniature-specimen mechanical testing technique scaled to the articulating surface of polyethylene components for total joint arthroplasty. *Journal of Biomedical Materials Research Part B, Applied Biomaterials* 48(1):75–81.

Lacalle, R., Alvarez, J. A., Cicero, S., and Gutiérrez-Solana, F. 2010. From archaeology to precious metals: Four applications of small punch test. *International Conference on Small Sample Testing Techniques* LXIII: 222–230. Ostrava, Czech Republic.

Murty, K. L., and Haggag, F. M. 1996. Characterization of strain-rate sensitivity of Sn-5%Sb solder using ABI testing. In *Microstructures and mechanical properties of aging materials II*, eds. P. K. Liaw, R. Viswanathan, K. L. Murty, D. Frear, and E. P. Simonen, 37–44. The Minerals, Metals & Materials Society.

Nomoto, R., Carricka, T. E., and McCabe, J. F. 2001. Suitability of a shear punch test for dental restorative materials. *Dental Materials* 17:415–421.

Parker, J. D., McMinn, A., and Foulds, J. R. 1989. Material sampling for the assessment of component integrity. In *Life assessment and life extension of power plant components*, eds. T. V. Narayanan et al., 171:223–230. ASME PVP.

Saunders, S. R. J., Banks, J. P., and Wright, M. 2001. Measurement of ductile brittle transition temperature of coatings using the small punch test. NPL report MA TC(A)60.

Stewart, G. R., Elwazri, A. M., Varano, R., Pokutylowicz, N., Yue, S., and Jonas, J. J. 2006. Shear punch testing of welded pipeline steel. *Materials Science and Engineering A* 420:115–121.

Tanemera, K. et al. 1988. Material degradation of long term service rotor. In *International Conference on Life Assessment and Extension*, 1:172–178. The Hague.

Villarraga, M. L., Kurtz, S. M., Herr, M. P., and Edidin, A. A. 2003. Multiaxial fatigue behavior of conventional and highly cross-linked UHMWPE during cyclic small punch testing. *Journal of Biomedical Materials Research* 66A(2):298–309.

Wanjara, P., Drew, R. A. L., and Yue, S. 2006. Application of small specimen testing technique for mechanical property assessment of discontinuously reinforced composites. *Materials Science and Technology* 22(1):61–71.

Zabihi, M., Toroghinejad, M. R., and Shafyei, A. 2014. Evaluating the mechanical behavior of hot rolled Al/alumina composite strips using shear punch test. *Materials Science and Engineering A* 618:490–495.

Zhang, L., Elwazri, A. M., Zimmerly, T., and Brochu, M. 2009. Shear punch testing and fracture toughness of bulk nanostructured silver. *Materials and Design* 30:1445–1450.

Index

A

ABAQUS, 103
Acoustic emission (AE), 37
AFM, *see* Atomic force microscopy
Ages named after materials, 1
Alloy design and development, 15
ANN models, *see* Artificial neural
        network models
Applications of small specimen testing,
        131–157
    coatings and surface-treated
            components, 144–147
        ball-indentation test method, 144
        ductile-brittle transition
                temperature, 144
        laser rapid manufacturing, 145
    condition monitoring of plant
            components, 131–132
        electric power industry, 131
        hot cells, 132
        nuclear industry, unique
                challenges in, 132
    electronic industry, 154
    field equipment for in situ testing,
            137
    material development programs,
            147–154
        ball-indentation-based data, 151
        biomaterials, 151–154
        composites, 149, 150
        face centered cubic Cu, 149
        metallic glass, 150–151
        nanomaterials and composites,
                147–150
        ultrahigh molecular weight
                polyethylene, 151
    pressurized heavy water reactor, 137
    residual life assessment, 138–141
        creep life, 139
        Cr–Mo steels, 138
        Vickers hardness testing, 138

sampling techniques, 133–137
    considerations in sample removal,
            136–137
    EDM sampler, 134
    fast breeder test reactor, 134
weld joints, properties of, 141–143
    disk specimen tests, 142
    fusion zone, 141
    heat-affected zones, 141
    small punch tests, 143
Artificial neural network (ANN)
        models, 64
Atomic force microscopy (AFM), 32

B

Ball indentation, 47–66
    analysis of pileup/sink-in, 59–61
    artificial neural network models, 64
    contact area and pileup/sink-in
            phenomena, 53–56
    cyclic indentation tests, 51–52
    elastic deflection, 54
    flow stress from indentation, 48–50
    friction effects, 58–59
    geometry of indentation, 47
    hardness tests, 47
    hidden layers, 65
    machine compliance, 56–58
    multilayer perceptron network, 65
    neural network, 65
    numerical methods for stress–strain
            evaluation, 61–66
    numerical studies on ball
            indentation, 56–66
    round-robin experiments of, 125–127
    strain definition, 50
    yield strength from indentation,
            52–53
Bhabha Atomic Research Center
        (BARC), 137
Biomaterials, 17, 151–154